MOTHER EARTH'S
Hassle-Free
Indoor Plant Book

MOTHER EARTH'S
Hassle-Free
Indoor Plant Book
by Lynn & Joel Rapp
Illustrated by Marvin Rubin

Published by J. P. Tarcher, Inc., Los Angeles

Distributed by Hawthorn Books, Inc., New York

This book is dedicated to Ruth and Bill, who created us—
and to Jeremy, who created *it* . . .

Copyright © 1973 by Lynn and Joel Rapp
Library of Congress Catalog Card No. 73-76660

ISBN 0-87477-007-6

Printed in the United States of America

Designed by Allan Franklin

Published by J. P. TARCHER, INC.
9110 Sunset Blvd., Los Angeles, California 90069

Published simultaneously in Canada by
Prentice-Hall of Canada, Ltd.,
1870 Birchmont Road, Scarborough, Ontario.

10   9   8   7   6   5

Table of Contents

I like plants.

I like plants a lot, in fact. So much so that when I get up in the morning the first thing I do after I pick up the paper and feed the fish and turn on the stereo is talk to my plants.

Sounds crazy?

Maybe, maybe not. I mean, let's look at the facts: I get up at 6:30 and everybody else in my house is still asleep. Okay, so who am I going to talk to until the rest of my family gets up? The paper?

"Good morning, paper! How do you feel this morning?" Ugh. Don't answer. I don't want to hear all that bad news.

Or "Hey, Stereo-baby! What's happening, man?" Not bad, I suppose. At least the stereo makes noise back. But somehow, still not real communication.

"Good morning, fish" is absolutely out of the question, especially when you've got a bunch of fish that look even more glassy-eyed than me.

But, "Good morning, Freddie!" Now that makes sense!

Who's Freddie? My Philodendron Panduraeforme, of course. And then there's Ruthie, my Dieffenbachia, and Basil, my Boston Fern, and Nedra, my Nephthytis. . . .

You might be interested to know how a fairly normal human being got so highly involved with a bunch of little green creatures.

It all began several years ago when I was pretty highly involved with another set of little green creatures— characters on a television show you might recall named "My Favorite Martian" for which I wrote several scripts. Now if you think watching television situation comedy is frustrating, imagine the terrors of having to write it! Brrr. I still shudder at the memory.

Anyway, one night I was driving home from a particularly nerve-shattering story conference wherein one of the lines in my script had been criticized on the basis that "A Martian simply wouldn't say that," and I decided I needed just a little nip to restore order to my head.

I pulled over to the curb, got out, and headed for a flashing BEER-WINE sign in the distance. But just then I spotted a little nursery and stopped to look through its foggy window. Something told me to open the door and go inside.

Good God! I'd stumbled into Paradise! The place was a

I've aiscovered my own Tahiti right in Hollywood

jungle of indoor plants—Palm Trees, Dieffenbachias, Philodendrons, Crotons. (Naturally, at this point in my plant
experience I didn't know these gorgeous things had names.
To me all plants were either Ferns or Palms, none of which
could live longer than an hour in your care unless you had a
degree in botany or horticulture.) Despite their anonymity,
they certainly presented a dazzling array of shapes and
colors and sizes. They smelled good. They looked beautiful.
The atmosphere was Serene.

Incredible, I thought! I've discovered my own Tahiti right
here in the center of Hollywood!

I started wandering through this glorious little rainforest,
and suddenly, the next thing I knew . . . I was talking to a
plant!

There I stood, spilling out the troubles of my day to some
poor, innocent, but uncomplaining Sansevieria Laurentii. (I
knew it was a Sansevieria—this particular plant was wearing a
name-tag.) What's more, it was listening! Unlike the average
bartender who puts up with a trouble-ridden monologue
because he has to, this silly-looking little plant really cared. I
could actually feel it sympathizing with every little moan.

But behind me, so was the nurseryman!

Properly embarrassed, I picked up the plant, bought it,
and took it home where I could continue my monologue.

The first thing I did was give it a new name. Tired of
calling it Sansevieria Laurentii (even "Laur" didn't ring
true), I named my new friend Irving, after a favorite uncle
who had a similar sort of stoic imperturbability.

I talked to Irving for another half-hour at least. Then I put
him to bed on my dresser and went to bed myself for the first
time in months without a sleeping pill. Quite by accident I
had discovered the perfect tranquilizer!

Now here I am, years later, not only talking to my plants,
but talking to you about them. From writer to plant-man to
writer-plant-man! Only in America!

How did it all happen?

Permit me one more flashback and I'll tell you, in
situation-comedy style, the story of The Birth of Mother
Earth just so all you other dreamers will see that dreams *can*
come true if you want them to badly enough.

FADE IN:

(Interior of TV writer's living room. A veritable jungle. Trees and plants everywhere. So much photosynthesis going on that the air is bright chlorophyll green. From somewhere amidst all the shrubbery comes his wife's voice.)

LYNN: Joel, do you realize there's no way we can live on what you earn as a TV writer anymore?

JOEL: Why not?

LYNN: I'll tell you why not! *(She appears from behind the fronds of a Palm Tree and points to a huge bush a few feet away.)* Last week that Schefflera alone ate $12 worth of dried fish-meal!

JOEL: Now, honey. . . .

LYNN: And swordfish-meal, yet! This morning I had to pay a plant doctor $15 to cure its mercury poisoning. Joel, I hate to tell you this, but these plants have got to go!

JOEL: Go? *(Starts getting emotional.)* Now just a minute, woman! Throw out the dog, get rid of the car, send the kids away, even. But if you so much as lay a hand on these plants . . . *(He begins moving around the room, crying at one plant, bending down reverently in front of another, stroking leaves, patting fronds. Tears begin to run down his cheeks.)* I love them more than life. I've raised them from budhood . . . nursed them through mealybug and root rot. I've trained them to stand tall in their pots and not wet the carpet. *(Wailing.)* How dare you suggest that we give them away?

LYNN: Give them away? I'm talking about opening up a store and selling them!
(Joel suddenly stops wailing.)

JOEL: Well why didn't you say so? *(He starts gathering plants into his arms.)* I'll load 'em in the car while you rent a location. . . . FADE OUT.

Exaggerated, I confess. But close.

It was because both Lynn and I felt so much better for having the plants, because we both knew how much better everyone else would feel for having them, and because we knew most people can't afford—or don't want to risk—

spending lots of money buying plants that are simply wrong, wrong, wrong. That's why we opened Mother Earth, and that's why this book.

We want to share our experience with you, and spare you, if we can, most of the heartaches we had to have and dollars we had to lose before we discovered the little secret truths about indoor plant care.

This houseplant book is different from any other house-plant book you will ever read. We make no claims to teach you everything. In fact, because this is a hassle-free book we've decided to leave out lots of things—but nothing we think is important for you to enjoy your plants and keep them alive. If you want to know how to take care of houseplants to the extent of being able to teach a class at UCLA or USC (or a Big Ten school, not to show sectional favoritism), then you should go to the library or purchase any of a great number of scientific works or "complete" books of indoor plant care.

At Mother Earth we are neither botanists nor certified horticulturists. We're just ordinary people who've had long and practical experience raising houseplants, and after having tried and failed and finally succeeded with virtually every plant that can be grown successfully indoors, we'd like to share it all with everybody.

So that's what this book is all about. Not just so much how to care for your plants—although we will be as thorough and honest about that as we possibly can—but how to become *involved* with your plants.

You may notice that this is one plant book which doesn't include pictures of the plants. Before anybody accuses our publisher of cheapness, let us tell you the decision not to include pictures was ours alone. It is based on our firm conviction that you have to see the plants—actually feel them around you—before you make your purchase. We do not want anybody looking at a picture of a plant, picking up the phone and telling his nurseryman, "Send me a Sanse-vieria," until he has actually seen the plant he is going to adopt. We can't say it often enough. We want you to become involved. Deeply, thoroughly, and emotionally involved, and as Lynn always says, "One plant is worth a thousand pictures."

And why should you become involved with your plants? I think Lynn said it best when she wrote:

A plant will make you happy. . . .
A plant will beautify your home. . . .
A plant will freshen the air. . . .
A plant will make a friend feel good. . . .
A plant will never talk back to you. . . .
A plant will never mess on your rugs. . . .
A plant will love you if you water it. . . .
A plant will give you something to talk about. . . .
And best of all—
YOU DON'T HAVE TO WALK A BEGONIA!

A plant will make you happy.

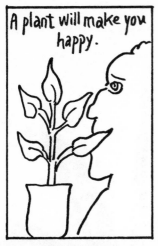

A plant will beautify your home

A plant will Freshen the air.

A plant will make a friend feel good.

A plant will never talk back to you.

A plant will never mess on your rugs.

A plant will love you if you water it.

A plant will give you something to talk about.

And best of all — you don't have to walk a begonia.

A green thumb is simply a positive state of mind.

A green thumb is simply a positive state of mind about growing things.

Raising houseplants successfully requires a combination of lots of actions—choosing the right plants for the right spots, watering properly, feeding properly, showing love, playing the right music—but most of all, having confidence!

You simply have to believe that those plants you've been watching die all these years can and will live in your home if you do right by them.

Every day I hear, "I can't make anything grow!"

"Hogwash," I reply. (I don't really say "Hogwash," but this is a family plant book.) "You can if you just start having a little confidence in yourself!"

"But how can I get confidence when everything I buy dies?"

"Buy this." I pick up a Sansevieria. "A Sansevieria was my first plant, and even though I didn't know a thing in those days, little Irving is still thriving. Fact is, he lived so well I had the courage to buy another plant, and then another. If he'd died . . ." I shrug my shoulders and sigh.

My point is, if the first plant you buy is something as hardy as a Sansevieria—often called a Snake Plant or Mother-in-Law's Tongue—the fact that it grows in spite of you can give your confidence just the boost that is absolutely essential.

But the next few plants you buy should also be hardy and easy to grow, or you may lose interest and give the whole thing up as a bad idea.

One other vital point: you're going to have to make a commitment. You'll have to be willing to give some time and some energy to the care of your plants. And after a while, as you see the results—the lush, healthy leaves, the new growth, even blooms, perhaps—you'll find yourself giving more and more of your time and enjoying every single minute of it.

I promise.

Mr. Mother Earth would never tell a little green lie! ᓄ

Chapter 1. The Green Thumb.

Buy your plants at a reputable nursery.

The most important factor in buying a plant for your home is to be absolutely sure that you buy an *indoor* plant.

Buying plants is not at all the same as buying clothes, or food, or automobiles. Probably the most important element that makes buying plants more chancy is the absence of a brand name. Say what you will, so-called brand-name products achieve that status only because of consistently good performance, a generally fair price, and consumer satisfaction.

The only way to compensate for this lack of brand-name plants is to buy your plants at a nursery—not a dime store, not a department store or discount house or supermarket or any of the hundreds of other places that are daily adding living plants to their inventory. Generally speaking, these people are merely jumping on the plant bandwagon and have no real interest in the plants or in you. Do buy your plants from someone whose business is selling plants. A specialist. A reputable nurseryman. Odds are you'll get healthy plants that will thrive in your home for years to come. (And just so the supermarkets won't get too mad at me, let me go on record here and now and strongly recommend that you never buy your groceries at a nursery!)

How do you know a reputable nursery?

First, it will carry a wide variety of premium-grade, healthy-looking plants. Second, it will have truly knowledgeable salespeople to help you. How to know they're knowledgeable, not being an expert yourself? Easy. If the salesperson knows all the names of the plants, if he's positive in his attitude, and (this is important) if he questions you about where you want to put the plant with regard to prospective light, etc., and then leads you to specific plants while sadly discouraging you from others along the way "because I'm afraid that plant would need more light than you can give it"—or whatever—then I would venture to say you're in a reputable nursery. In fact, I'd bet on it!

Next, make sure the plant you're buying is an indoor plant. (If its name is listed in the Index of this book, you can be certain it is. But look up what we say about it because not all are easy to take care of.) Another indication that it's an indoor plant is if you find it in either a plastic or terra-cotta (clay) pot displayed inside a room or a greenhouse. If the plant is in a metal can and making its home outside or in a

Chapter 2.
What an
Indoor Plant Is
and How
To Buy One.

lath-house, consider it an outdoor plant.

When you feel confident that the plant you want is in fact an indoor plant, your next step is to try to make sure it's healthy and of the highest possible quality.

You'll find it's easier to tell a healthy plant from a sick one than it is to judge its quality. A sick plant will have droopy leaves, yellow or brown spots, sparse foliage. Make sure the plant you buy is a rich green color, and that the leaves are free of excessive discoloration. (Plants such as the Dracaenas carry brown tips on their leaves as a natural accessory. But again, if you are dealing with a reputable nurseryman, he'll tell you whether discoloration is natural.)

Check the plant for signs of disease such as mealybug (a fuzzy white cottony-looking substance), red spiders (microscopic mites that suck chlorophyll out of leaves so that the leaves of a sick plant appear pale), or scale (rows of hard shiny bead-like insects generally attached to stems and leaves). (More on bugs and plant diseases in Chapter 5.)

Be sure your plant is in a good potting mix, which, if you're shopping at a reputable nursery it most always will be. Briefly, indoor potting soil is always more porous than just plain dirt, and should contain small white rock, such as pieces of vermiculite.

Now, once you've determined to the best of your ability that the particular plant is healthy and properly potted, try to find out if it's truly a premium-quality plant.

This part isn't easy if you're on your own. Plants are very much like eggs when it comes to judging quality. Put a Grade A Large Egg next to a Grade AA Large Egg and see whether you can tell the difference simply by looking at it. You can't, and neither can I. Put a Grade B Dieffenbachia next to a Grade AA Dieffenbachia and I doubt very much whether you can tell the difference in that case, either. Learning how to differentiate between real quality plants and lesser quality plants takes years of experience.

But there is one helpful clue: price.

For the most part, indoor plants are grown in greenhouses, and the heat of the greenhouse is the major influence on the growth rate of the plants. The hotter the greenhouse, the faster the plant will grow, thus the cheaper it will become to the consumer. But, alas, the less chance it will have to survive the transition from the torrid humidity of its

Make sure the plant you are buying is an indoor plant.

birthplace to the natural, generally drier atmosphere of a home.

Plants grown with care and attention in a cool greenhouse naturally take a great deal longer to attain marketable size. Thus, because time is money, a premium-quality plant will always be more expensive than a second-grade plant, even though the two plants may look exactly alike.

(Writing of the way plants look reminds me of an incident that happened not too long ago. My wife and I had made an appearance on a Los Angeles television show, and had, as is our custom, brought along a few of our plants, all of which have first names. The next day a woman came into the store and asked to see Alvin, a stunning Chinese Evergreen who'd been on the show. When Lynn brought Alvin out, the woman gushed, "My goodness, he's even more handsome in person!")

In any event, remember that you'll generally find no real bargains in houseplants. You get just what you pay for.

Naturally, the larger the plant, the longer it has taken to grow and, as a result, it will be more expensive than a plant in a smaller size. However, presuming you intend to place at least one or two large plants in your home, I would recommend you purchase them first, that is, after you've gained confidence from your Sansevieria.

Don't say, "Oh, no! What if they die? I'd rather start with a small plant and see what happens." If you buy a plant according to your light situation (on which I'll shed more light later) and if you get proper care instructions, there's no reason why you should have any problem. In fact, the larger the plant, the easier for you! It's just like people—the young ones take more care and consideration, while the older ones demand less of your time.

Okay, stop twiddling those thumbs which are already beginning to show the first hints of verdancy. Let's move to Mother Earth's GTGL—Guaranteed-To-Grow- List!

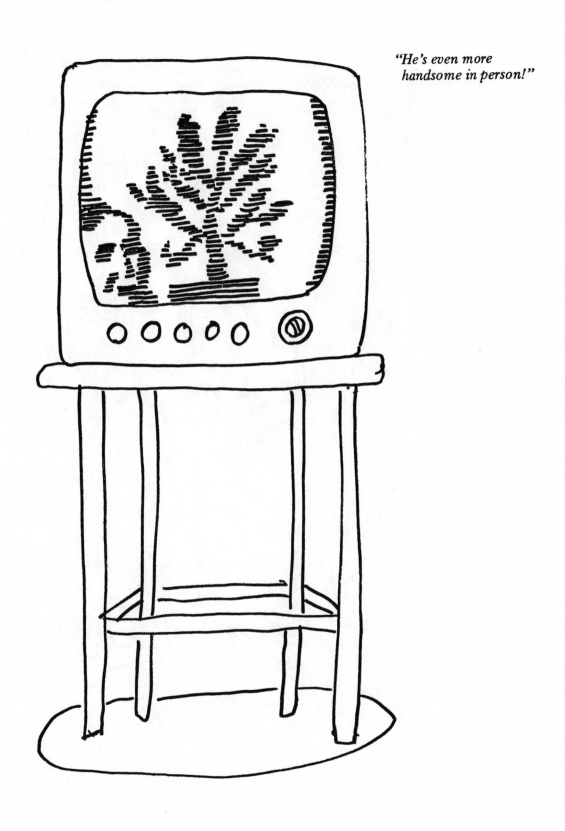

"He's even more handsome in person!"

In Victorian England no home was complete without an Aspidistra.

Before we begin, one important caution. The successful raising of indoor plants is not an exact science. Plants are living things and simply refuse to act in a totally predictable fashion. As circumstances vary, so will their life patterns. Water quality and climate change from place to place. Even the air is different in various locales, and procedures that work in Maine may not in Texas. All plants will, under certain circumstances, exercise their God-given right and fail in spite of the fact that you have followed our directions to keep them growing. Don't give up on yourself or Mother Earth just because you have lost a plant. Try again, secure in the knowledge that if you keep growing forward down your own particular garden path, you and your plants will win a lot more often than you will lose.

There are basically three categories of indoor plants:

1. Easy to grow.
2. Hard to grow.
3. Impossible to grow.

Unless you're some kind of masochist, I would imagine that you'd like to begin your plant experience with the easy, almost impossible-to-kill group.

Because Lynn and I have had such good luck with all the fifty-six plants in our living room (you read right—56, count 'em, 56), I find it just a bit difficult to segregate a small list of the easiest plants without hurting the feelings of all the rest. But, secure in the knowledge that all my plants love me and will understand, here are fourteen basic plants that even Jack the Ripper or the Boston Fern Strangler couldn't do in. . . . Don't let the names scare you. You'll probably mispronounce most of them at first, but these days how many people speak fluent Latin anyway? Perhaps the safest thing to do would be to write down the names of the plants you want to buy—or take this book along with you—and then let your nurseryman show you to your plants.

Aspidistra—The Aspidistra was probably the first "houseplant" as we know them today. In Victorian England no home was complete without an Aspidistra in the parlor. However, because Aspidistras seem to have fallen out of popular favor, nice ones may be difficult to find. If you *should* come across an Aspidistra, grab it! One variety has shiny black-green leaves; one has green and white striped leaves. Usually broad, the leaves can grow to 2 feet in length.

Chapter 3.
Fourteen Basic Plants (and a Few of Their Relatives) Guaranteed To Grow For You and One That May Not.

The plant will grow any place—light or dark—and without much care. Water when it feels on the dry side and feed it once a month. Mist the leaves—we'll tell you how later—a couple of times a week. Final proof of this plant's hardiness is its common, common name, Cast-Iron Plant. Good enough?

Boston Fern—Everyone loves Ferns, but some people feel they will never be able to make a Fern live. When we talk about Ferns, we are usually referring to the Boston Fern (Nephrolepis Exaltata "Bostoniense") and her beautiful relatives the Roosevelt, "Fluffy Ruffles" and Whitmani.

Give your Boston Fern a home in a Fern stand or hang it from a beam or a hook in the ceiling. Keep it off those shelves and tables and specially off the floor. The fronds must fall freely—not touching walls or tables—so that air can circulate all around the plant. Ferns need medium light, nothing exceptional, but will fade and burn if the direct sun gets to them. A cool, well-ventilated room is best. Hot stuffy rooms are out, and, as with all plants, keep them away from a heat source.

Ferns love water, and in a standard terra-cotta pot, they're almost impossible to overwater. Spray with a fine mist every chance you get. Water *thoroughly* once a week (so that the water runs through the bottom of the pot at least three times). Every couple of days, water about an ounce per inch of pot size. Always water by lifting the fronds and pouring water directly into the soil. (If you water into the crown, it may not dry out due to the thickness, and the center of the Fern will rot.) On your weekly watering day, take your Fern someplace—either to the sink, the bathtub, or outside—and after you water him, give him a good haircut. Snip out all the brown fronds, which are inevitable both because the Fern is reacclimatizing to your house from the greenhouse and because old fronds are always dying to make room for new ones.

If you feed your Fern the proper amount of Spoonit once every two weeks, and if you love it a lot the way we love ours, you'll become a member of a very proud fraternity— The We-Can-Make-Ferns-Grow Society. (Confession: We have so many Ferns growing in our house now and they all look so beautifully alike, we've had to paste their names on their pots so we can tell one from the other.)

Chinese Evergreen (Aglaonema Modestum and A. Simplex)—They are fantastic. Strictly for indoors, they like to be watered when dryish and have a broad light spectrum—low-medium light to good indirect light. Variegated members need a little more light than the darker green ones. (*Variegated* means that the leaves come with patterns on them and in different colors, as opposed to solid-colored leaves.) Occasionally a leaf or two will turn yellow, but don't let that alarm you. Snip the yellow leaf off down at the base of the stem with ordinary scissors. If a great many need snipping, you may be overwatering. In addition to Aglaonema Modestum and A. Simplex, the species includes Marantafolia, Treubii, Pseudo-Bracteatum, and others. Many Aglaonemas bloom little lily-like flowers and produce colorful berries from time to time. A relatively carefree plant, they have a long life span and can grow to a very large size—as tall as 4 feet.

Dieffenbachia—An old standby, the Dieffenbachia comes in many varieties—Amoena, Exotica, Ostredia, Rudolph Roehrs, Superba, etc. Leaves vary in color, markings and size. The dark green leaves of the Amoena can reach 18 inches in length, while the variegated white, yellow and green leaves of the charming Exotica are quite small and pointed. All do well in good indirect light, preferably close to a window. Water when plant is dryish. It is important not to overwater this one, especially when the soil is cool. It loves warmth. Here's the one plant you can buy when it's very small and watch it grow into overwhelming proportions. It's a pretty plant to look at, but don't put it in a salad! The juice of the Dieffenbachia will make you temporarily mute. Hence the name—Dumb Cane! I know one thing for sure—they're not deaf. We talk to ours all the time. She answers us by shaking her leaves. She gets really happy when we mist her—especially in warm weather. Jackie, our Dieffenbachia, was one of the first plants we owned. She's doing quite well, thank you, for a lady of about twelve years (almost 100 in human terms), especially since we started talking to her in Greek.

Dracaenas—Dracaenas are the perfect houseplants because they thrive in a low-medium to high-medium light range—anything but the dark or the direct sun. Water them when they feel dryish, whether they are 5 inches tall in a

4-inch pot or in a 24-inch tub and towering 10 feet above your head. Don't overwater. And please, don't be tempted to clean their beautiful shiny leaves with any leaf-shine product or olive oil or mayonnaise or milk. All these things have oil in them and should not be used because oil clogs the stomata (breathing cells) and will suffocate your plant. When the leaves need cleaning, use Basic H or Amway LOC, biodegradable cleaners (2 teaspoons per quart of warm water). Sponge leaves and wipe with a soft bathtowel. Gently (always). Feed the small plants once a month, and the large trees three times a year.

The most popular Dracaenas are the Dracaena Marginata, recognizable by its ultra-thin dark green leaves bordered in dark red; Dracaena Massangeana, whose broad green leaves have a striking chartreuse stripe down the middle; and Dracaena Warneckii, with its green and white leaves.

All Dracaenas are stalk plants. That is to say, as they grow, they lose their lower leaves, so don't worry when yours do.

Ficus—Among the members of this exciting family of houseplants, some, like Ficus Decora (Rubber Tree), are commonly known. Ficus Lyrata (Fiddle-Leaf Fig), often also called Ficus Pandurata, is another member of the tribe that enjoys living indoors. Ficus Benjamina (Weeping Fig) is the star of the family because it looks exactly like a tree you might find growing in your yard (it looks very much like the birch tree), and yet will thrive beautifully in your home.

General care for these is basically the same. The Rubber Tree is the strongest and can live happily in a medium-light situation. Other members need what we call good indirect light and will grow best near a window. Let them get fairly dry between waterings and, of course, spray them every day.

Ralphie, the Ficus Benjamina in our living room, has been with us for more than two years. He goes through stages when his leaves turn yellow and fall off, but no reason to panic. New leaves form where the old ones were. Trimming off what appear to be dead branches is a NO-NO. When you purchase a Ficus Benjamina, be sure that it has been "adjusted" to indoor living by being in a greenhouse for at least three months. The Ficus Pandurata is tough. Its leaves are broad and flat and resemble elephant ears. These must be kept clean. If leaves begin to fall off, you may be overwatering.

A Dieffenbachia is one plant you can buy when it's very small and watch it grow to overwhelming proportions.

Grape Ivy (Cissus Rhombifolia) and *Kangaroo Vine* (Cissus Antarctica)—Great indoor plants, Grape Ivy and Kangaroo Vines need good indirect light and can grow nicely under electric lights. Don't overwater, but mist daily. If it's in a plastic or glazed pot, water only when dryish, which may be every two weeks. If it's in a standard clay pot, water more often, but only when soil feels dryish 1 inch below surface. We have an 8-inch Grape Ivy in a plastic pot that gets watered thoroughly every two weeks. Then she almost dries out before her next drink. Grape Ivy grows fast. You can keep it full by pinching off new growth (use thumb and forefinger, not scissors), but we like to see how long ours can grow. It should be pinched back if it gets "leggy" or sparse in the center. Feed monthly.

Maranta (Prayer Plants)—Mother Nature has provided Marantas with a natural cup: When the leaves close up at sundown, they are actually holding moisture in their little praying hands. In their natural tropical habitat this moisture is provided for them, but in your home you are the someone they pray to for their daily misting with warm water and for whatever humidity you can provide for them. Without that the leaves turn brown on the tips and eventually dry up, so make sure the soil is always moist and set the plants on dry wells (see pages 52-53). They look better if pinched back and shaped, and they can also be trained up bark or moss.

Nephthytis (or Syngonium)—Nephthytis will grow almost anywhere indoors. It will even grow well in water. Leaves are arrow-shaped, with the exception of the green Nephthytis; most are variegated. Caring for these reliable and fun plants is easy. Water when the soil is beginning to feel dry, but don't let it get too dry. The plant will tell you when you are overwatering by turning yellow. Nephthytis grow quickly and no two grow in the same way. They love to be *crowded* in their pots and adapt well to decorative containers.

Palms—Some of the popular varieties available in today's plant market are the Phoenix Robelleni (Miniature Date Palm), Neanthe Bella (Dwarf Palm), Kentia, and Chamaedorea Costaricana. The latter two are large Palms, perfect for home decorating. The many members of the Chamaedorea clan need medium to good indirect light. The reliable indoor varieties all need uniformly moist soil. The Kentia Palm is fed by spraying a food with iron chelate onto the plant, not

We bathe our plants every Saturday night whether they need it or not.

into the soil, as it's been our experience that the root ball structure of the Kentia does not absorb the plant food successfully, and soil feeding over a long period usually does harm to the plant. Spray with water daily. All Palms are hardy and will love you dearly if you bathe them. Sponge off the fronds with a mild mixture of biodegradable cleaner and water, then rinse off with plain water every three to six months. Actually a good way to keep larger Palms clean is to take them into the shower with you. (As for our Kentia, Ricky, we give him a bath every Saturday night whether he needs it or not.)

Philodendron—We can't possibly list and describe all the Philodendrons that can grow indoors because there are just too many. They are vines and in their natural habitat grow across the floors of the jungle and up through the trees, so in your home they're happiest when they're climbing up stakes or trailing out of their pots. Put one of these reliable plants in a decorative container and it takes on a whole new personality. One of our Hastatums makes its home in a marvelous old tin from the 30s. Counting our plants, we find four Cordatums, one Rubrum, two Hastatums, one Wendimbi, one Niostera, one Panduraeforme, one Cannifolium—all Philodendrons and all thriving and happy.

Don't be tempted to shine the leaves of your Philodendron with anything but biodegradable cleaner and water. Water plants when soil begins to feel dry 1 inch below surface. Spray daily with a mist of warm water and if the plant is growing on a stake, be sure to spray the stake, too. The plants can be fed by the usual method or by spraying food onto the leaves, but don't feed too much as they are fast growers.

Incidentally, we're a little upset by the fact that many people "put down" Philodendrons. "Oh, they're so common," we hear all the time. Sure, they're common. But does that mean they're not beautiful and entitled to equal rights? In an effort to see justice done to the Philodendrons all over the world, we have formed a group known as the NAACP—The National Association for the Advancement of Common Philodendrons. If you wish to join, just do so by saying these secret words to your nurseryman next time you happen by: "Two Philodendrons, please."

Piggyback (Tolmiea)—This is the one plant on our list that

Don't feed your Philodendrons too much as they are fast growers.

is sometimes temperamental. We are including it on this list because its lush green fullness makes it one of the most popular plants at our shop and most of our friends have had good luck with it. The Piggyback got its nickname because little plantlets grow on the backs of more mature leaves. Although resembling a geranium, it does not bloom. It's effective decoratively because it is full, plump and lush. Piggybacks do well in good indirect light. Coolness and good air circulation are important. Water enough to keep soil uniformly moist—about one cup every other day when in a terra-cotta pot. Do not mist, but do keep plant on a dry well (see pages 52-53). Piggybacks like a sponging with Basic H and warm water every couple of months to help keep mealybug and red spider away, but no daily misting because of their hairy leaves.

If your Piggyback looks all droopy, just relax. Take it outside and give it a little drink. In about an hour or two, it should be standing tall, good as new. Or almost. (Many Piggyback owners go through this trauma, and it pains us to think how many perfectly good Piggybacks have gone to Plant Heaven when all they needed was a good strong drink.) If the plant begins to turn brown around leaf edges and lose its color, put it outdoors—any place but in direct sunlight—water daily and watch what happens. It comes back, big and strong! But then it should be left outdoors, except in cold winters with frost, when it should be brought back inside and put in the sunniest spot in the house.

Pothos (Scindapsus)—In reality a vine, Pothos is everybody's first plant. (At least it's the other first plant besides Sansevieria.) The leaves are variegated. Two main varieties—Golden Queen and Silver (or Marble) Queen—can grow many feet long. (Would you believe 15 feet?) Many think it's a Philodendron, but it isn't, although it is similar in appearance and equally hardy. Easy to care for, it can tolerate any light situation except extremes—no dark corners and no direct sun—and even does well in containers with no drainage hole. Must be kept indoors. Water when soil begins to feel dry.

Sansevieria—That's him. Our old friend Irving. Our first plant and, as we said earlier, still thriving away, sitting on our mantelpiece in the living room. Commonly called Snake Plant, probably because of its shape and color (usually tall

Sansevieria is the most hassle-free plant.

and stiff and variegated with yellow markings), it doesn't really look like a snake to those of us who love it. Sansevieria can survive in bright light or low light or anywhere in between. It can withstand almost total neglect, but if you treat it as it should be treated, the plant will appreciate it. A Succulent, it stores a good deal of water in its leaves, making it unnecessary to water very often. Probably the only way you could commit Sansevieria-cide would be to overwater it, but we suggest you don't try.

Spathiphyllum—Yes, it is difficult to pronounce, but easy to love. The "Spath" is very hardy, can live in low to very good light, blooms a lily-like flower off and on over the year and is quite lovely. This is a particularly good plant for those of you who like to water, as it needs to be kept quite damp. As with most indoor plants, Spaths love humidity and will adore it if given a daily misting. Setting them on dry wells (see pages 52-53) will prevent their leaves from drying up. Feed bi-monthly. Loves to eat.

You will note that we've left out such familiar and popular names as Aphelandra (Zebra Plant), Azalea, Begonia, Croton, Gloxinia and Maidenhair Fern.

Those plants will grow in your house, all right, but they'd grow better if you lived in a greenhouse. We call some of them Glamour Plants and tell you more about them in Chapter 12.

Onward to Basic Principles— ᗧ

*Spathiphyllum is
particularly good
for those who love
to water.*

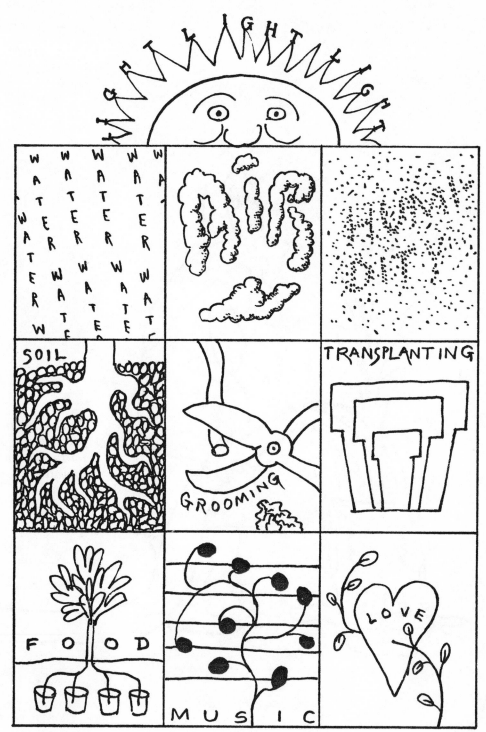

Basic principles of plant care.

Now that you've bought one or several of those Guaranteed-To-Grow plants, let's talk about the generalities of houseplant care. In essence, the amount and quality of your involvement.

All plants need ten basic things to be happy in your home: Light, Water, Air, Humidity, Food (Fertilizing) and Soil, Grooming, Transplanting (occasionally)—and Music and Love.

That's right, Music and Love. Scientific experiments have proved that the right music can have an enormous effect on the growth rate of your plants. And the need of all living things for love is a universal truth. It is well known that plants grow best to classical music, but we have been told about a hip Dieffenbachia who loves The Rolling Stones. It's all a matter of taste. But if you are away from your plants all day, some music should be played for them. They languish without life around them, so keep your plants where *you* live—they want to be with *you*.

But also important are things like Light, Water, etc.

Light

All houseplants need some light, but very few, if any, need or even want direct sunlight. Especially sunlight magnified through glass. Most plants in that circumstance will melt faster than a popsicle.

Keep plants away from direct contact with glass windows or doors. Glass is a conductor for heat or cold, and drafts are a danger, too. The leaves must not become cold, and if they get too hot, their function, photosynthesis, breaks down. (See *Webster's Unabridged Dictionary*.)

As a general rule: Most houseplants will do just fine with average filtered light, either natural or artificial. As we progress, we will deal with specifics, such as what plants will do well in minimum light and which plants must have maximum light in order to thrive. But generally speaking, if your plant is not getting enough light, it will begin to wilt, turn pale, and die. If it's getting too much light it will shrivel, turn yellow, and die.

Filtered or diffused light means light coming through a window, filtered by light curtains, adjustable shutters, venetian blinds, shades, trees, even screens on windows. We don't refer to indoor light as "sunlight." We always refer to it as "filtered light." Of all the light sources, it is the best for your plants. Almost every houseplant needs filtered or diffused light and almost none require direct sunlight, or can live in it.

Most people have the opposite problem—not enough light. If that's true for you, you can rearrange your plants every week, moving the plants from low light to medium light (never from a low-light situation into a bright sunny window). They aren't crazy about being moved around, but if you love plants around you, and your house or apartment lacks sufficient light, then you have to do anything you can to make things better for your green friends.

Wherever they are, plants always grow toward the source of light. And be sure to turn them frequently so that growth will be even all the way around.

If there is no natural light available and you're a plant lover, don't despair. Artificial light will get the job done. (Growing plants under lights is becoming increasingly popular. Some people are even growing plants in their closets, although I often wonder what kind of plants a person would want to grow in a closet, and why.)

Incandescent Light

Amazingly enough, many plants will respond to ordinary incandescent light (that's the kind you have in ordinary light bulbs). As long as the plant is fairly close to the light (but not too close, of course—at least 12 inches away from the bulb), and the light remains on at least 12 to 16 hours a day, most indoor plants will do okay (with the same exceptions listed below for fluorescent lights). However, while fluorescent lamps provide cool light that does not dry out the air, incandescent bulbs give off heat, which does dry out the air, so plants must be kept on moist pebbles or dry wells (see pages 52-53) and sprayed daily if possible.

Fluorescent Lights

Many modern office buildings have offices without windows. The only light comes from fluorescent light fixtures, which causes plant-loving office-dwellers no end of concern. But they should *not* be concerned, for a great many plants will do beautifully with only fluorescent light. Among the best are Dwarf Palms, Dracaenas, Chinese Evergreens, Philodendrons, Sansevierias, and Nephthytis. Ferns don't do particularly well under fluorescent light, and certainly none of the more exotic, colorful plants such as Crotons or Aphelandras can be expected to sustain in these circumstances. But don't despair, where there's fluorescent light, there's hope.

Similar to fluorescent, but more sophisticated, specially developed Plant-Gro lights actually do a better job than natural daylight in many instances, particularly in the culture of blooming plants such as African Violets and Bromeliads. They are a Westinghouse product, so if your local hardware store doesn't stock them, you now know whom to call.

As we pointed out earlier, there isn't a houseplant alive who likes to sit in a blazing south window where the sun pours in, magnified through the glass.

EXCEPT:

Cactus!

And what's wrong with that? You'd be surprised how sensational a Cactus garden or an arrangement of Cacti can look, but since Cactus is a true specialty and entire books have been written on the subject, we'll just say that if you have a spot that would melt an ordinary houseplant, don't hesitate to play around with our prickly little friends. They come in dozens of shapes and sizes and they even bloom.

If we move just a bit away from the window now into an area that gets lots of extra-good light but no direct light for too long a time (say, not more than two hours a day), quite a few plants will do just beautifully, among them:

African Violets	Kangaroo Vine
Ficus Benjamina (Weeping Fig)	Piggybacks
Fittonia	Schefflera
Grape Ivy	Succulents
Ivy	

· We've already talked about most of these, so now let's take a moment to make green-thumbnail sketches of the four new acquaintances—all recommended for those extra-good light situations:

African Violet—The only blooming plant that's truly a houseplant is the African Violet. (Although some Spathiphyllums produce lovely white spathes, they're not blooms in the true sense.) If you keep African Violets in extremely good light, water them frequently with tepid water, and feed them a special African Violet Food weekly, they'll bloom like mad all year round. Just one other thing about getting African Violets to bloom: It doesn't hurt to pray a lot.

Fittonia (Nerve Plant)—A table plant with delicately veined green or pink variegated leaves, the Fittonia needs good light and enough water so that it never dries out. It will droop when it gets too dry. Make a dry-well arrangement (as described on pages 52-53) for your Fittonias because they love humidity. Mist lightly two to three times a week with warm water. These spreading plants tend to get leggy but look better when fat and full, so when they start crawling all over the table, pinch back.

Ivy—This is the name given to several varieties of vine, but its real name is Hedera. There are many, many ivies—English Ivy, Glacier Ivy, California Ivy, German Ivy—so it goes. We recommend leaving most ivy outdoors. If you do try it, give it a home in very good light. Water generously and spray daily. If it starts to yellow and lose leaves, put it outside in indirect light (a little morning sun or late afternoon sun won't hurt it a bit) and water frequently. Easily grows into a lush, full and large hanging plant.

Schefflera—Commonly called an Umbrella Tree, the Schefflera is easily recognized by its large green five-leaf-clover leaves. Scheffleras are around in great number today, from small ones in 4-inch pots to huge ones in tubs. They are relatively hardy and can live indoors for many years. They need good light (very near a window is recommended) and water only when they get quite dry. This is one plant whose soil must be allowed to get almost bone dry between waterings. However, a daily misting with warm water is a must. Keep the plant clean, as Schefflera is the red spider's filet mignon. It's terribly important to give your Schefflera a bath or shower whenever possible, especially the undersides of the leaves. A cool airy room is vital.

Lots of other so-called indoor plants are always listed as plants that do well in extra-good light. Among the more common of these are the Anthurium, Azalea, Cyclamen, Begonia, and Maidenhair Fern. And just about all of the other blooming plants—Gloxinia, Chrysanthemums, etc. The problem is, none of these plants will really live indoors in the light. They'll just live longer than if they were in a dark spot in your home. Quite honestly, they have been popularly sold through the years as houseplants but are, in fact, much better suited to culture outdoors (see Chapter 12).

I recommend leaving most ivy outdoors.

We don't recommend you buy them unless you can afford the luxury of replacing them periodically or have become very expert in plant care.

Plants for the Dark

"What can I put on a bookcase that gets no light at all?" Lucille Ball asked one day in the store.

"Hold on a minute, Lucy," I said. (When I was writing her TV show I called her Miss Ball. See how plants can change a person's life?) "Maybe there's more light in that corner than you think."

"Now YOU hold on a minute, Mr. Rapp," said Lucy (when I was writing the show she called me Joel). "I know light from dark, and—"

I put up a reassuring hand. "I know. But you'd be surprised at some of the things that will grow in a spot most people are sure is too dark. There's Spathiphyllum, and Neanthe Bella Palm, and—"

"Joel!" Her voice was firm and authoritative. "I've tried Spathawhatever, and I've tried that other one, too. If you must know, I tried a plastic plant in that spot and even *it* died!" She stared at me with a triumphant gleam in her eye.

I swallowed, then said bravely, "How do you feel about mushrooms?"

"Mushrooms?" Her eyes narrowed.

"Why not?" I babbled. "I'll admit, a mushroom isn't the most attractive plant in the world, but if it's really all that dark, who's going to see it anyway?"

Needless to say, Lucy didn't go for the mushrooms (as I recall, she went for my throat), but she did agree to think about installing some kind of artificial light above the bookcase, the only possible solution in a case as extreme as that.

But how can you tell if that dark corner where you'd love to have a plant is actually too dark for anything? Well, you can try to cast a shadow on the spot during peak light hours. If you can, it's safe to say *some* plants will live there. But then if you try one of those listed below, and you find after a few days that it's getting saggy, or turning black, move it and give up on gardening in that spot.

Recommended night people are:

Our recommended night people are:

Aspidistra	Nephthytis
Baby's Tears	Philodendron
Chinese Evergreen	Sansevieria
Dwarf Palm	Spathiphyllum

We've already discussed in some detail all the plants listed above except Baby's Tears, and since this is a gorgeous, easy-to-care-for plant, let's have a go at it right about here.

Baby's Tears (Helxine Solierolii) looks like lush green carpeting in a pot and will grow in little clover-like clusters, both tall and flowing over the sides. It's easy to grow in the house and is good in terrariums. Water when it feels slightly dry to the touch, feed monthly, and spray with warm water only two or three times a week. This plant is a snap to propagate. Just cut about a 1-inch square out of your pot and transfer it to a smaller pot, roots and all. We wouldn't advise taking it from outdoors, where it is also grown as groundcover, because ordinary outdoor dirt is not sterilized, and bacteria and creepy-crawlies are brought in along with the plant.

Other experts may include many more plants in this low-light category. We've seen recommendations for Fittonia, Dieffenbachia, English Ivy, Boston Fern, and Schefflera, among others. Frankly, though, our own practical experience has taught us that *those* plants—as well as *any* plant with vivid color or blooms—simply will *not* grow in a dark corner.

I suppose what I'm doing is blatantly asking you to ignore what you've heard or read elsewhere and pay attention only to me. That takes nerve, but I ask it from the heart. For your dark corners, use only the plants I've listed. And remember, you heard it from the man who recommended mushrooms for Lucille Ball's bookcase!

Water All plants need water to live. Certain plants, like our old standby the Sansevieria, need very little water. Other plants, called *Succulents,* need even less, as they store their water in their leaves. But all plants, including Cacti, need water. And being stuck in pots in our houses, they depend on us to give it to them.

Watering is probably the most important part of plant

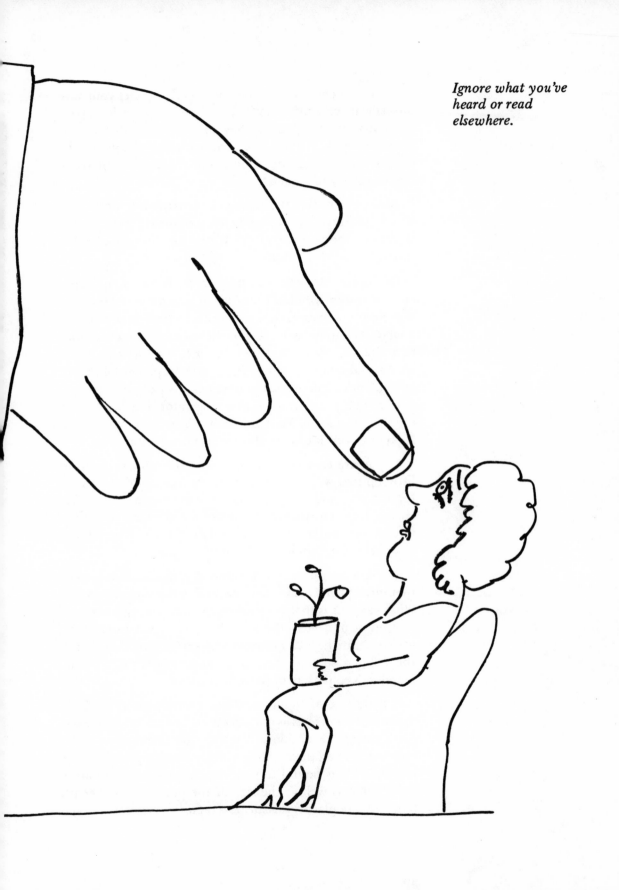

Ignore what you've heard or read elsewhere.

care. Too much—death. Too little—death. So feel your way through it carefully, until you find the schedule your particular plants need. If they look droopy, they're not getting enough. If their leaves start turning yellow, or the plant is droopy or if soil is muddy or the stems begin to rot, they're getting too much.

Always make sure the water you use is room-temperature, about 70 degrees. That's *warm* if out of the tap. If bottled, it is already room-temperature. Water run through a water-softener is definitely out.

Of course, the best possible water to use is rainwater. Mother Nature wouldn't use it if it weren't perfect, would she? So if you happen to live in a place where it rains once in a while, try to collect a few bottles of clean rainwater (not rainwater off the roof). If you live in Southern California as we do, and rain is just something you hear people talk about in fond, nostalgic tones, the next best type of water is water you've defrosted from your refrigerator and brought to room temperature. Most impurities have been frozen out. Next best is unfluoridated bottled drinking water.

If you are restricted to tap water, we can give you one little tip that will benefit both the plants and you: Cover the top of the soil with about an eighth-inch of purified charcoal, which will strain some of the chemical impurities out of the water, as well as purify the air. Bags of this charcoal can be purchased at almost any nursery.

We don't suppose it will come as a big surprise to learn that more plants die from overwatering than underwatering. What happens is that your average plant-owner gets worried that his plant needs a drink, so he waters it and waters it and waters it until he either drowns it by suffocating the roots, or the soil gets so cold that the plant gets root rot, the plant kingdom's equivalent of pneumonia.

A good rule of thumb is to water your plants just about the time the soil feels as if it's dry about 1 inch below the top (with certain exceptions, Crotons, Aphelandras and Ferns foremost among them). How quickly this happens since the last watering is bound to vary depending on conditions in your home as well as the size of the pot and whether it's terra-cotta or plastic. (Plastic pots retain moisture longer and thus the plants don't need to be watered as often, which is an advantage. Yet your plant can't breathe quite as well in plastic as in terra-cotta, which is a disadvantage.)

One way to find out if a plant in a terra-cotta pot needs water is to tap a table knife on the side of the pot. If the sound is like a ring, the plant needs a drink. If the sound is a dull thud, the soil is still moist.

Check all plants in 6-inch and larger pots weekly (smaller pots should be checked more often). When you are first becoming acquainted with your plants, let your finger be your guide. Feel the soil about 1 inch below the surface. If your plant is in commercial potting mix, a very porous substance in which just about every indoor plant you buy will be (except for certain Palms), you will have no difficulty getting your finger in up to the first knuckle. If the soil feels dry, then a thorough watering is necessary. The simplest way to do this is to take the plant to the kitchen sink, fill the pot up to the brim with warm water and allow the water to run through the root system and out the bottom of the pot. Take the pot out from under the tap until the water has stopped leaking heavily and then put it under again, once more allowing the water from the tap to run through and out the bottom. Do this one more time—three times in all. This is called watering-through.

If the plant is too large to carry, give it enough water, a little at a time, so that you can see water coming out of the bottom of the pot into the dry well on which we hope it is resting (see dry well, page 53). Never allow the plant to sit in the water. This causes root rot and the damage is usually irreparable.

If you cannot get your finger into the soil, then you have to use a somewhat more hit and miss method. If the soil is dry and flaky and resists even the gentle proddings of a fork, the chances are your plant needs water. How much will depend on the size of the plant and how quickly it drinks what you give it. If the water that you give it (slowly) sits on top of the soil for more than five or ten seconds, you can safely presume that it now has had enough to drink. If the water is absorbed immediately and air bubbles up from the soil, your plant is thirsty and probably needs one more for the road.

47

Remember that the little plants need looking after and may even need to be watered a little every day. If you have plants that like to be kept uniformly moist (which does not mean wet), then you will have to make sure they never get dryish. This means watering more often, but not deep-root waterings, possibly just an 8-ounce glass at a time. This all varies with the size of the plants.

We have certain large plants that need a good drink about every three weeks. This is unusual, but we point it out because it is really remarkable how long plants can survive with minimum water in their pots.

Naturally, the more light your plant gets, the sooner it will dry out and the more often you'll have to water it. So it's safe to say your plants in the dark corners will see you and your watering pail a lot less often than your plants in the brighter light. But stop by on your rounds and say hello. Just so they won't get the feeling you're ignoring them.

There are mechanical devices available that will help you determine when your plant needs water, or some very sophisticated devices that will even water your plants for you. As tempted as you may be to purchase one of these modern-day miracles, please don't. Remember the whole secret of success with plants is your own personal involvement, and any self-respecting plant will resent these mechanical intrusions, not to mention your lack of total concern. If you find such mysticism hard to believe—we're sorry. Because it's true.

Be cautious in watering, don't overdo. Check plants regularly. Have a relationship with your plants, and before you know it, they will be talking to you and you will understand them.

Air   Plants need air, just as we do. Both their roots and their leaves. In her plant classes Lynn always lists a fork as one of the things you need to care for your plants.

Don't get nervous. We're not suggesting you have to eat your Philodendrons. Just use that fork to aerate the soil about once a month, especially if the plant's in a plastic pot. Those roots have to get air, and if the soil gets too hard and no life-giving oxygen can get through, it's curtains. But do be careful to poke the fork only about a half-inch into the soil in order not to damage the roots, and always work from the

The young ones take more care.

outer rim toward the center. Root band-aids are not only expensive, but difficult to find.

As far as the leaves are concerned, they need air in order to breathe. And as you probably know, if plants can't breathe, neither can you! Plants suck up the carbon monoxide in the air and convert it to oxygen which they then spill out into our atmosphere. Of course, they reverse this procedure, but (and how smart can you get?) they do it at night, when most everybody is sleeping and not too much oxygen is being taken in. An explanation of this Wonder of Nature is not to be found in a hassle-free plant book like this one.

There is a distinct difference, however, between seeing that a plant gets air—that is, making sure it isn't stuffed in a cabinet or an enclosure—and putting it in a draft, which is definitely bad for the plant. Make sure your plant isn't in front of an open window (unless it's a very mild day) or by a door that is opened and closed frequently.

We're constantly being asked about the effect of air-conditioning on indoor plants. Our feeling is that plants do better in an air-conditioned room than in a blazing hot one. Don't you? Naturally it would be a mistake to put your plants directly in the stream of the air-conditioner, but the constant temperature of an air-conditioned room is much better than the rising and falling peaks and valleys most of our climate zones experience. Obviously the major draw-back to air-conditioned rooms is lack of humidity. But with just a little effort you can compensate for that. How? Read on.

Humidity   Humidity is crucially important to every houseplant. Remember, most of these plants grow wild in tropical climates, and the closer we can come to approximating their natural habitat the better they'll grow.

A major killer of plants is dry heat, and you can fight it with a greenhouse or with a spray bottle, the latter being infinitely less expensive and infinitely more practical, to say the least.

Spray your plants with a fine mist of warm water every day. Please. They'll survive if you don't but they'll love it (and you) if you do. We don't recommend using a Windex-type bottle because the effort required can quickly turn a green thumb into a black and blue one. A plastic spray

Plants do better in an air-conditioned room. Don't you?

bottle, **available** in most nurseries or beauty supply shops for about $2, is best. We hope you get one with an adjustable head on it so you can create a fine mist. If you really want to get fancy, plant sprayers in metal, ranging from aluminum to platinum, are available in many boutiques and specialty shops. But these are more for decor.

Spraying is fun and it's virtually impossible to overdo. Stand approximately 6 inches away from your plants with your spray bottle in your hand. Spray until your plants are covered with a fine mist. It's a good idea to spray the undersides of the leaves periodically because this is where little critters hide, like the mealybug and red spider, and you are helping keep the plants clean. Also this is where the plant ingests moisture.

Grouping plants is another ideal way to increase humidity. And occasionally you can take them into the bathroom or laundry room for a humidity-bath.

(Speaking of baths, if you're planning a trip and you can't find a plant-sitter in your neighborhood, don't worry much. In Chapter 7 we tell you how to make a greenhouse out of your bathtub.)

During the summer, if you live where it gets blistering hot, take your plants outside in the shade and spray them liberally. They'll love that. And you. We promise.

A final tip on humidity: If possible always set your plants on "dry wells." Cover the bottom of a tray or saucer with a layer of pebbles. Set your plant on top of the pebbles and add water to the tray or saucer so that the water level is just below the bottom of the pot. Although the base of the pot doesn't touch the water, evaporation beneath the pot provides humidity. It's best to use a saucer or tray that is glazed on the inside, and one-quarter to one-half inch of pebbles. Put enough water in the well every day to keep pebbles covered. Water evaporates quickly. Plant must never sit in water, just on moist pebbles only.

Food (Fertilizing) and Soil

Plants have to eat, too. But in moderation.

We really hate to get pedantic, but here, briefly, is how the whole thing works:

A good houseplant always comes potted in a special soil, rich in the nutrients necessary for its growth indoors, and sterilized to prevent growth of fungus and bacteria and to get

Dry well.

Water level is just below the bottom of the pot.

53

rid of stray seeds which might sprout and really confuse the issue.

(Yes, it's quite possible to make up your own houseplant soil. However, since this is a hassle-free book and since there is probably no greater hassle in the world, outside of baking a cake from scratch, than making up houseplant soil from scratch, we avoid telling you how to do it. Let's face it, in the twentieth century certain things made from scratch are strictly impractical.) It's much easier to buy a bag of good sterilized indoor potting mix.

But back to that plant in its pot. It stands to reason, doesn't it, that sooner or later those hungry little roots will eat up all the food that's in that little pot of soil? So about once a month you should replenish that nutrition.

Now we've all heard a great deal of discussion about acid-loving plants and alkaline-loving plants. Many people recommend that "acid-loving plants" be fed with special acid-loving plant food. Other people feel that certain plants that prefer alkaline soil should be potted in alkaline soil and fed with a more alkaline feed. However, the truth of the matter is that this is a very sophisticated type of indoor gardening. We believe that the plants that we recommend— the plants that will live the longest in your house—are all plants that will do just fine in regular, store-bought potting mix, and fed with the all-purpose foods that we talk about.

Essentially what you need in your plants' pantry are 1) plant food which might be either an organic fish emulsion or one of a number of chemically created indoor plant fertilizers and 2) a vitamin hormone supplement. We'll tell about the vitamin hormone use in the section on transplanting and when we get to caring for sick plants.

Whatever food you choose be sure to read the manufacturer's directions carefully. Read them at least three times. Believe them. Too much food can burn the roots, and most fertilizer is very powerful. If you overfeed your plant by using too much too often, the plant suffers terribly—all the way to the garbage can. So be careful.

If you are using a fish emulsion, the soil must be dampened before the fertilizer is added. For this method, pre-water the plant, either the day before or in the morning approximately three hours before feeding. (Watering and

Plants have to eat too.

feeding and misting should always be done in the morning. And sparingly!)

If you are using a chemical-type plant food, it can be applied as part of the watering. Make sure all plant food is thoroughly dissolved in warm water and properly measured. Use the solution of food and water as the last run-through for your plants.

Fast-growing plants and flowering plants (Ferns, Piggybacks, African Violets) can be fed very diluted solutions once a week (or if you're lazy, twice a month is okay, too).

Plants on boards and stakes can be fed through the soil, but it's a good idea to spray dissolved plant food on the large leaves of these climbing plants and on the leaves of large Dracaenas. Plants are capable of ingesting food through their leaves; this is called foliar feeding. Kentia Palms should be fed only through the leaves.

Lots of people put tea on their plants, or old coffee grounds, but our own experience tells us it's better not to treat our plants like begging dogs. I mean, coffee grounds probably won't hurt most plants, but there are plants that are allergic to acid and caffeine, and it might even keep them awake all night—so why take chances?

Keep another thing in mind when it comes to feeding. Like all other living things, plants need a rest every once in a while, too. When you notice one of your plants has stopped growing—gone dormant, as it is said—hold up on the feeding until the plant wakes up and starts growing again.

All plants should be given a two-month rest period from all food. We recommend December and January because this prepares the plants for their growth period in the spring.

Also you should never feed a sick plant (check table, pages 76-77, on symptoms). Any time you suspect a plant is sick, be aware that plant food of any kind will have a damaging rather than a healthful effect on the weakened root system. Remember, you must always feed sparingly at best, since the slightest overfeeding can cause irreparable harm. Be sure to discard "leftovers." Food mixture cannot be saved from feeding to feeding.

In summary, be sure you only feed your plants a little bit and only on a regular schedule. Don't be a Jewish Mother to your plants. It's a good idea to keep a calendar or diary so

*Don't be a
Jewish Mother
to your plants*

you'll know when to fertilize. It isn't possible to remember when a month has gone by. Please don't trust your memory. Write it down. Have a schedule, especially when you're just beginning. (This goes for watering the plants, too.)

Grooming   Not so mystical, this aspect, since it consists of cleaning and cutting. Doesn't it make sense that if you take a little trouble to wash off the leaves of your plants from time to time—remembering never to use anything but plain warm water or a mixture of water and biodegradable cleaner—and trim out any brown or yellow leaves—your plants will feel better, look better, and thus grow better?

Of course it does.

As mentioned earlier, many plants develop brown tips from lack of humidity or from chemicalized hard water. (Spraying with bottled [unfluoridated] water can help retard this.) Because the plant's energy goes into the green healthy part, not into dead or half-dead leaves, those leaves will never be green. Plants with a tendency to develop brown tips, such as Dracaenas and Marantas, should be trimmed with sharp scissors at least once every two weeks.

Don't worry about hurting the tips of the leaves when you start snipping. It doesn't hurt when the barber cuts your hair, does it? (It would, however, if he pulled your hair out by the roots—so don't yank off dead leaves. Always snip them.)

When you're grooming, just be careful not to cut too far into the green part of the leaves. Trim off the dead edges to the same contour as the leaf.

Ferns need their hair cut more often than children. It is inevitable, with Ferns, that old fronds will die off, as new ones are constantly reappearing. So if you're sure that your Fern is getting plenty of water and humidity, don't fret when you see those brown or yellow fronds. Bring it to the sink, take your scissors, and for a few glorious minutes, become the Vidal Sassoon of the plant set.

But remember what we said earlier about leaf-shine. Don't finish up your plant's haircut by applying any polish.

Just like in real life, in the plant kingdom greasy kid stuff is passé.

*Ferns need their
hair cut more
often than children.*

Transplanting   One of the most feared, yet one of the easiest aspects of indoor gardening is the common transplant operation.

No need to call in a horticultural Dr. Barnard for this simple task. Just remember a few basic rules and in most cases your plant will come through the operation good as new. Better, in fact, since the object of a transplant is to allow the roots new freedom to grow.

However, almost all houseplants like to be very close in their pots. They enjoy close quarters, so don't transplant unless it's necessary—maybe the plant has outgrown its pot or hasn't grown for years or had mealybug and recovered or is going into a pretty new-pot home.

How do you know when your plants need to be transplanted?

The best real indicator is the sight of roots growing out of the bottom of the pot. Sometimes, however, if there is no drainage hole or the roots are packed so tight they can't grow out the bottom hole, it will be necessary for you to remove your plant from its pot and check the roots.

Lift the pot, turn it upside down, and lightly tap one of its edges against a table or a sink. The plant should slide right out, its roots and dirt contoured to the pot structure. If the roots are obviously overcrowded—that is, if you can see tons of roots and just a little dirt—it's time to put that plant into a larger pot.

Step 1: Get a larger pot, usually one that is 2 inches wider in diameter than the present pot.

Step 2: Place a clay pot chip over the hole in your new pot so the water won't run through TOO fast. Or, if you're planting in a container without drainage holes, be sure to fill the container at least one-fourth of the way up with broken-up clay pot chips or pebbles and charcoal. This will absorb and evaporate water so your plant won't get root rot.

Step 3: Put in about an inch of potting soil.

Step 4: Now loosen the dirt around the roots of the plant. Be gentle—but don't be afraid. And work quickly so you don't expose the roots to the air any longer than absolutely necessary. By removing the dirt by a gentle massage at the bottom of the roots, and extending them, you'll be giving them a huge headstart into their new growth pattern. And remember, the quicker the roots grow, the faster they'll be ready to send up new energy to the plant. (In

60

case we haven't mentioned it before, it's important to understand that if you put a plant into a pot that's TOO big, either the roots will be all that grow, or they will spread out and fall apart, and the foliage will remain very small and in time, may wither up and D-I-E.)

Step 5: Put the plant into its new container, adding or subtracting dirt at the bottom until the plant is at its desired decorative height. Now start adding sterilized potting soil around the plant, packing it down tightly until it's all level about a half-inch below the top of the pot.

Water thoroughly, either with warm water or a mixture of SUPERthrive and water (see pages 73-74). Allow the water to drain through at least twice. And voila, the transplant is complete!

Every now and then, a plant will fail to react well to the transplant and will go into what is known as "transplant shock." If your recent transplant begins to droop soon after the operation, suspect shock at once, run to your nearest nursery and buy a bottle of SUPERthrive. Apply as directed, only as directed, and within days, the patient will recover.

At least usually.

It isn't possible for one person to transplant a tall tree or large plant by themselves, and it isn't always necessary, unless of course the plant has outgrown its pot. (Then find a friend fast.) But if the plant has been sitting in the same soil for a few years, you might want to give it a little "help" by "topping the soil," a very painless type of operation. This is how it works.

Topping the Soil

Have a friend hold the tree so balance is not upset, scoop out as much of the old soil as you can without disturbing the root section. It's easier to do when soil is slightly moist. Make sure you have plenty of newspaper on the floor—it can get messy. When scooping out the old soil, always work from the outer rim of the pot toward the center, to protect the roots. Now replace the old soil that you have removed (and thrown away) with fresh sterilized potting soil. Pack it down as firmly as possible (in plant talk this is called "tamping"). Leave a few inches at the top of the pot (usually the width of the rim) so that water will be absorbed and won't run over the sides. Run-through either with warm water or a mixture of SUPERthrive, which should prevent any shock. Your

61

Mother Earth's Hassle-Free Transplanting Chart

1. Transplant if the roots have grown out of the pot

2. or examine roots. Turn over & tap lightly against a firm surface.

5. Loosen dirt around roots & gently extend them.

6. Set on dirt. Add dirt until plant is proper height.

3. Use a pot about 2" larger in diameter.

4. Put a broken clay chip over the hole.

If there's no hole – fill ¼ pot with clay chips or pebbles and charcoal. Add 1" potting soil.

7. Fill & tamp. Fill & tamp. fill & tamp.. up to ½" from edge.

8. Let warm water drain through twice.

plant will appreciate this kind gesture and will probably sprout a few new leaves or grow a few inches in a very short time just to show you how much! Be sure to aerate weekly.

Music and Love

It has been written that music soothes the savage breast. It also helps the gentle little plants to grow.

This, my friends, is an authenticated, scientific fact.

Let's have a look at some of the work done by Les Harsten, a 28-year-old sound engineer from New York. Les has analyzed how plants respond to sound, and in an experiment using two banana plants—each getting identical light, warmth and water, but one being exposed for just an hour a day to a special high-pitched hum—he discovered that the plant receiving the hum had grown faster and was actually 70 percent taller!

According to Harsten, the sound he developed stimulates a plant's breathing cells, or stomata, to stay open for longer periods of time. Thus it takes in more nutrients than it normally would and the plant grows faster.

(If the sound is played continuously, however, the stomata never close and the plant grows so fast it kills itself. Plantacide!)

A special record incorporating Harsten's sound is now on the market, and it's being used by several commercial growers. But to tell you the truth, I think Brahms or Beethoven works just as well.

Studies by the Smithsonian Institute and the U. S. Department of Agriculture indicate that plants have definite classical leanings. They prefer peaceful to dissonant music. Hard rock can kill them.

The scientific fact seems to be that what the plant responds to—whether music or voices—is the vibratory energy: air moving faster than normal around the plant. Any vibratory energy will stimulate growth.

Now that you know that talking to the plants is not crazy, now that we've shown you that it's scientifically important, start talking without embarrassment or fear.

Meanwhile, back at the lab—

Cleve Backster, a New York polygraph expert, decided to hook a pair of electrodes to a plant. Surprised to see that the reaction pattern was similar to a human pattern, he dipped a leaf in hot coffee to see if it would react. Nothing happened.

Plants have definite classical leanings.

So he decided to burn the leaf. He left the room to get a match, and on his return he noticed the tracing on the graph had already made a jagged upward leap. Such a leap on a person's tracing would mean fear and anxiety. Just Backster's intention of harming it seemed to have triggered a response in the plant.

In another test conducted by Backster, six people went out of the room and drew lots to decide who would kill a plant. They reentered one by one. The polygraph showed a sharp upward jag when the plant-killer entered.

Backster's research hasn't convinced everybody that plants have emotions, but some scientists are now duplicating Backster's research using, instead of a polygraph, an ultra-linear oscillograph. These people hope to discover whether plants actually have the ability to think and be aware for themselves or whether they are only mirroring the thoughts and feelings of the plant-handler.

One foundation spent over $2,500 last year just for such research. They could have asked us; we'd have told them the answer for nothing.

We believe. We love our plants and they grow. We talk to them, and we're sure they can hear us. But we knew before the tests. A plant is dumb; it doesn't have any vocal cords; it doesn't reason. But it can respond. Sure, it's a miracle, but it's a miracle we're all here. Once you believe that, you have to believe anything. ᕙᕤ

Just the intent of harming it seemed to trigger a response in the plant.

Diseases & Pests.

As people in the plant business, take it from us: The worst pest when it comes to killing plants is Homo sapiens.

It's true. Overwatering, overfeeding, or complete neglect by *us* cause a great many more plant deaths than insects do. However, those little buggers DO take their toll, so it's best you know who they are and how to do them in.

Remember, first and foremost, you must check your plant thoroughly before purchase to make sure it is free of disease. If the plant looks healthy, the leaves are all lush and green, and you see no signs of insects or diseases, okay—take it home.

But please remember—common pests such as *mealybug, red spider, aphids* and *scale* are in the air and are attracted to plants, so don't expect your nurseryman to refund your money or replace your plant if it should happen to contract some disease. However, although the diseases are common their occurrence is not, so don't lose any sleep over it.

It has been said that an ounce of prevention is worth a pound of cure—and in the plant kingdom, that goes double or triple.

How can you go about preventing the sudden appearance of mealybug or red spider?

The best advice we can offer is (1) to check your plants every few days, talking to them as you go, and make sure you see no telltale signs of disease and (2) to give your plants a thorough washing about once every two weeks. That's right. Take them to the sink, tub, or outside in the shade and wash them off. Red spiders and mealybugs often hide under leaf joints and in hard to get to places, but are easily dislodged, so a good washing down will generally get rid of developing problems. Nip them in the bud, so to speak.

If, however, your best efforts go astray, let us assure you that you probably can save your houseplants from just about any disease *if* you catch it in time.

Mealybug, one of the most common houseplant diseases, is a white, furry-looking inanimate insect that gathers in white, furry-looking clusters on your plant. (Inanimate insect here means one that doesn't move—not one that's dead.) You can't miss it. It's white and furry-looking and a bit sticky, too. How to get rid of it?

If you detect mealybug in the early, early stages, it can be attacked with a Q-Tip dipped in rubbing alcohol. Dip the cotton swab directly into the alcohol and then swab down the plant where the mealybug colonies have taken up residence. Spray off with warm water next day. This is advised for small plants only, as it takes a lot of time and patience to attack and conquer. You can also take regular rubbing alcohol, mix about a 50-50 spray with water, and spray it on your infected plants.

If the disease is clearly visible and more advanced, try the water from the shower, hose or spray on the sink. A small toothbrush comes in very handy for those hard-to-reach places.

If this fails, you can spray an insecticide called malathion on the leaves or place it in the soil. Malathion can be a real killer, so be very, very careful not to use more than prescribed on the bottle. Even a little less is a good idea. Always use any spray outdoors, and don't use malathion more than once a week. Spray plants that were close to the "sicky," too; then throw away any leftover mixture. Keep all insecticides out of the reach of children—in fact, keep all your plant supplies in a "safe" place.

If you suspect your plant has a disease, always isolate it. Mealybug is catching. Keep it isolated until you feel confident you have cured it.

If the plant recovers, give it a good bath in a Basic H and water solution and put it back in its place. When you are sure that mealybug is gone, transplant into fresh soil, as mealybug lays her eggs in the soil. But a *word of warning:* mealybugs are sneaky. They come back. So for the rest of its life, keep your eye on that plant.

If all this is to no avail and mealybug was too far advanced, say the final rites and bring in a replacement.

Red spider is a much more difficult problem. Unlike mealybug, you can't see the virtually invisible little red spiders, or mites, as they are often called. But you can see their damage. Leaves turn silver, as the chlorophyll disappears, or turn yellow as they die. Leaves fall on the floor for no apparent reason. A slight red dust can be seen on the underside of the leaves. That's the red spider. Get a magnifying glass and you can easily see them move about. The monthly washing is a good deterrent, but if your plant gets red spider, malathion or mite spray is your only hope. Prepare yourself emotionally for a small, green death in the family. Red spider is almost impossible to cure.

Aphids and *scale* are the other two most common houseplant pests. (Moving plants from inside to outside to inside is a good way to bring aphids inside. This is not advised unless the plants are carefully examined.) Aphids are easily recognizable as small, winged insects that affix themselves to the leaves. As small as they are, it isn't uncommon to hear one aphid say to another, "I can't believe I ate the WHOLE thing!" So be rid of aphids with a good dousing of malathion.

Scale is actually the thickened and hardened skeleton that protects the insect in its adult stage and also its eggs, which the female lays inside the scale before she dies. Found almost exclusively on Ferns, scale is the most difficult plant pest to destroy. Malathion or Black Leaf 40 (nicotine sulphate solution) is your only hope.

Several other types of insects prey on plants, but it is safe to say that it's more important for you to be aware that your plant is under attack than by what kind of army. As long as you can determine that a failing plant is the victim of either a predator or a disease—rather than dying at your OWN hands or perhaps just from natural causes—the best bet is malathion or some other spray specifically labeled indoor plant spray—always used to the manufacturer's specifications.

(With regard to plants just up and dying for no apparent reason, we've got to accept the idea that it happens in the plant kingdom just as it happens in our own. When it happens—and if you have enough plants it surely will from time to time—try not to shed too many tears, and then put the plant to rest in the garbage posthaste.)

Sometimes fungus or other undesirable elements can come from the soil, rather than the surrounding air. That's why your plant must be potted in sterilized soil.

Just remember that pests and fungus and even death "come with the territory" in the cultivation of houseplants, so try not to be overemotional about your plants.

I remember one woman who really loved her plants and every time there was a trace of mealybug or the slightest hint of red spider, she'd call up, panicked, wailing "What can I do?"

We tried to be as tolerant and helpful as possible, but one night she went too far. About 1 o'clock in the morning on a particularly windy night, the phone rang in our bedroom, and sleepily I picked it up. It was our lady.

"Joel," she practically screamed, "a big gust of wind just shook the house so bad my Boston Fern fell down and came out of its pot! What shall I do?"

Fighting my anger, and filled with sleep, I yawned and replied, "Give it two fish tabs and call me in the morning."

We've got to accept the idea that plants just up and die for no apparent reason.

Every once in a while, a plant will get "sick." Sometimes its illness is caused by overwatering. Its leaves yellow and drop off. The plant droops. Sometimes a plant is situated where it cannot get enough light. Its leaves turn pale, or black splotches appear in the center of the leaves. The plant droops. Funny as it may sound, sometimes the symptoms for overwatering and underwatering are quite alike. Only you know what you have been doing to your plants . . . you have to be the doctor.

Once you have ascertained that the plant is ailing, put it in the hospital. Your hospital can be anywhere near a light source where you have extra space, maybe in your laundry room. We found that the top of the clothes dryer was a marvelous place for reviving tired plants because of the humidity in there.

If the weather is mild, set the patient outside, but only in the shade. We find it very difficult to throw a plant out, so we always have a little hospital zone and it's truly amazing how plants can recover from overwatering, overfeeding, and all kinds of things. Not always. And it does take time and energy, but they can revive . . . and that's a good feeling—for you *and* the plant.

A good treatment for a sick plant is the application of a wonderful product we have found. It's a vitamin-hormone supplement called SUPERthrive. This is not a food. It is a food supplement, similar to vitamins for us. As root-toner and conditioner, it helps healthy plants grow and does all kinds of miraculous things. One word of warning: use sparingly. Even less than that.

The recipe for the use of this wonder-drug is this:

First Application (as part of the watering), 10 drops per gallon of water.

Next Application (when plant has to be watered again), *one* drop per gallon. (That's right. One drop per gallon. Use the cap or buy an eyedropper from your druggist. I think an eyedropper costs 17 cents, and it's a good investment.)

From that time on, use it EVERY OTHER WATERING until you see marked improvement in your plant. If the plant doesn't respond after two or three months, it's time for the plant to meet its maker.

SUPERthrive can be used when transplanting, as the

Chapter 6.
The Plant Hospital
or What To
Do Until the
Doctor Comes.

watering-through that the newly planted need. It can also be used on your healthy plants, but we advise this only during their growth period, which for most plants, indoor and out, is spring and summer. If you have any left over from your indoor plants, you can pour it on your outdoor plants, but it must not be saved from week to week.

Any insecticide, food, or plant product that has been mixed must be used that day and then thrown away.

A word of warning: Please keep all plant food, supplies and insecticide out of the reach of children and animals.

I love to play young Dr. Kildare to my plants when they are under the weather. They appreciate it so, and usually perk up in no time.

For handy reference use the chart shown on pages 76-77. Although any chart of this kind is a gross simplification of what's wrong, why it's wrong and what can be done to fix it, you need some guidelines, and the information given here will at least start you on your internship in plant doctoring.

*I love to play
young Dr. Kildare.*

Mother Earth's Hassle-Free Remedy Chart.

Symptom (What does it look like?)	Cause (How did it happen?)	Cure (What can I do?)
Yellowing leaves. Soft stems. Wilted leaves. Dropping leaves.	Too much water.	Allow plant to dry out. Trim off yellow leaves. If possible, put out-doors in shade until revival.
Lower leaves drop off (especially Dief-fenbachias). Plants have no luster. Not growing well. Leaves dry and crackling.	Not enough humidity.	Group plants. Mist frequently. Set on dry wells (see Humidity, pp. 50, 52).
Plant droops. Doesn't grow. Lower leaves yellow. Foliage dull.	Not enough water.	Immerse in warm water 30 sec. (once only). In-crease frequency of watering or increase amount of water (but not both, not too much).
Plant usually dies, or new growth grows in blackened and misshapen. Has tummy-ache.	Overfed.	One "flushing" of pot (one immersion in warm water for 30 sec.).
Plant lacks luster. Color fades. Doesn't grow. Leaves yellowing, turning brown on tips and getting small and smaller.	Underfed.	Begin regular regime of fertili-zation. Start with half-strength first feeding. For brown tips feed with iron (plant has tired blood).

Symptom (What does it look like?)	Cause (How did it happen?)	Cure (What can I do?)
Plant wilts frequently. Fades or burns.	Too much light.	Move into medium-light situation.
Leaves fall off. Plant gets leggy. Droops. The Worst!	Not enough light.	Move into medium light. Three days later move into good light.
Black spots on leaves. (Plant has cold.)	Draft.	Move plant out of draft or send it abroad for 3 months.
White fuzz on leaves, stem and on soil.	Mealybug.	Early observation. Immediate treatment. Malathion.
Web-like matter on undersides of leaves. Visual proof: A stranger on your plant.	Red spider and other pests.	Patience. Prayer. (See Chapter 5.)
Plant looks sad. Shakes. Afraid. Has migraines.	Pressures and loneliness.	Play nice music for plants. Be confident. Get to know your plants better. Don't worry too much about them. Relax! Read and learn as much as you can. Plants are very understanding.

Back in two weeks, sweetheart.

Chapter 7.
Vacation Care.

When you go on vacation, you will always worry about your plants. People ask us so often, "What can I do about my plants while I'm away?" Here are a few solutions.

The best thing you can do is hire a plant-sitter. However, if such a service is not available or convenient for you, train a friend, or perhaps you're lucky enough to know someone who knows plants, who has the green thumb. If so, take the time to make a chart of your plants showing where they're located, and note how much water they get and how often. Introduce the sitter to the plants and let her (or him) watch while you water them. Leave the feeding for when you get back. Chances are the plants will miss your special touch (and they do get used to it), but when you return they will be alive.

It helps if you can put all the plants in a centralized location where they are easy to take care of—and, of course, where there is light. Make sure some fresh air can get into the house while you're away.

If a plant-sitter is out of the question, round up all the plants and take them into the bathroom. Fill the tub with water about 6 inches deep. Invert terra-cotta pots and set your plants on top of the pots. Then take a large plastic bag from the cleaners and use masking tape to tape it to the walls around the tub or over the top of the tub. Leave the bathroom light on while you're gone. The plants will thrive in the greenhouse you've created for them, but don't stay away longer than a couple of weeks.

Another bag method can be used also. Simply stick bamboo sticks into the pots after they have been thoroughly watered. Drop a "baggie" over the top of each and fasten around the pot with masking tape. Set all plants on a tray (cake pan or pie tin) filled with pebbles. Add enough water so plants are sitting on moistened pebbles. This method is good for seven days.

Another alternative is separate vacations—or maybe just staying home in your own jungle and pretending you're in Hawaii. ᔆᓬ

Nathaniel Ward is generally credited as the inventor of the terrarium but the Chinese were planting in glass hundreds of years before.

For days I had been struggling with my conscience. I had a problem I couldn't solve, an answer I couldn't find. What, I looked Heavenward for help, should I title my chapter on bottle gardens?

Terrariums, which is the common, un-snooty term?

Or *terraria,* which is the preference of my good strict constructionist friends?

At any rate, I finally decided to cop out, because what the heck, a terrarium is, after all, a bottle garden, isn't it? At least, most of the time it is. (Actually, it's any group of plants put together in a glass container—a fishbowl, a brandy snifter, an aquarium. Historically, the inventor of the terrarium is generally credited to be Nathaniel Ward, a London physician who lived some hundred years ago. He devised a glass box called the "Wardian Case" for shipping plants. But the fact is the Chinese were planting gardens in glass hundreds of years before.)

Terrariums—or terraria—are absolutely wonderful. Not only are they beautiful and decorative, but they're probably the lowest-maintenance living thing you can own. A few drops of water once a month and you're home free.

Where do you get a terrarium? Well, you can buy one almost anywhere these days—nurseries, plant shops, markets, shoe stores—and you can spend a good deal of money, too, depending on the size of the container. Or you can make your own.

It's something you'll dig doing.

That is, do digging.

All you'll need to start with are these basic items: planting tools, bottle, soil, charcoal, pebbles, crushed-up clay pots, plants and patience.

Almost any kind of bottle will do. We usually plant our larger terrariums in five-gallon water bottles and our smaller ones in one-gallon wine bottles. (Emptying the wine bottle will provide a measure more pleasure, I suppose, but filling up either kind with plants is a ball.)

Most people, when they see the little bottles full of plants, think we start them from seed and they grow in the bottle.

Wrong.

We plant our bottle gardens with small plants—usually ones that come in 2-inch pots, but some larger terrariums will take 4-inch plants—i.e., plants that come in 4-inch pots.

How to make a terrarium.

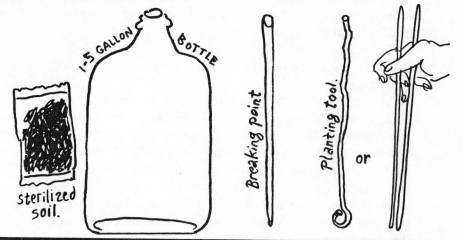

sterilized soil.

1-5 GALLON BOTTLE

Breaking point

Planting tool.

or

1. Empty bottle.

2. Twist stiff paper or aluminum foil into a cone.

3. Add pebbles.

5/6

1/6

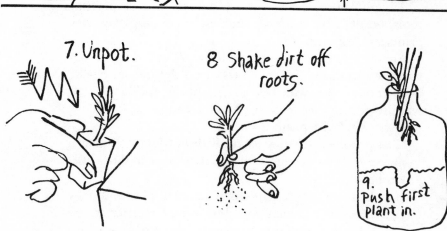

7. Unpot.

8 Shake dirt off roots.

9. Push first plant in.

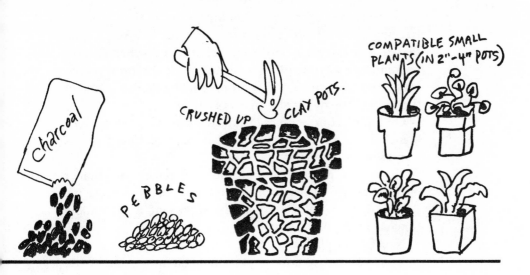

Charcoal

PEBBLES

CRUSHED UP CLAY POTS.

COMPATIBLE SMALL PLANTS (IN 2"-4" POTS)

4. Add charcoal, then soil.

¼"

5. Make a planting scheme.

6. Make a 2" small hole in the soil

10. Bring to proper angle.

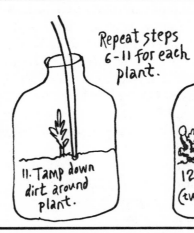

11. Tamp down dirt around plant.

Repeat steps 6-11 for each plant.

12. Add frosting- (twigs, stones, etc.)

To begin, using some type of thin funnel that will reach to the bottom of a thoroughly dry bottle (make the funnel by twisting a piece of stiff paper or aluminum foil into a cone shape), put a layer of ordinary pebbles or crushed-up clay pots into the bottle at least one inch thick. This will absorb water and prevent humidity from building up too high and the plants from dying of root rot or fungus.

On top of your pebbles or crushed clay pieces, sprinkle a ¼-inch layer of charcoal. You can buy charcoal for this purpose at your nursery.

Next, put in your soil. About 1½-inches for smaller bottles and 2½-inches for larger ones. Again, make sure it's sterilized, because it's especially important not to allow fungus to form in the bottle.

Now for the plants.

You can put almost any small plant into your bottle garden, but the ones that seem to flourish best are Variegated Table Ferns (Pteris), Maidenhair Ferns, Baby's Tears, Dwarf Palms, Dwarf Dracaenas, and Lycopodium Moss. (For color, you can try a small Croton or a bright Ti Plant, but for the most part there's just not enough air or light for a flowering plant in a bottle garden.)

Variegated Table Ferns resemble carrot tops except they have white running through the fronds. Lycopodium Moss grows in clumps, is chartreuse, looks like crepe paper and dies quite fast.

It's important to be sure that the plants you mix are compatible. That is, not only must they be able to get along with one another, but they should be plants that require the same amount of water and light. Makes sense when you think about it, of course, but you'd be surprised how many people don't think about it. You can create a handsome, compatible family with little Dracaenas, Sansevieria, Philodendron, and Nephthytis, for instance. Another nice group consists of Maidenhair Fern, Dwarf Palm and Baby's Tears. Ferns and Moss make a perfect base material. So if you're planning on mixing up a plant dish, be sure you've got the right recipe.

Figure out your planting scheme on a table top before you start placing the plants in the bottle.

For planting, you'll need a long, thin tool of some kind.

And also, a long, thin breaking point. For in the beginning, planting a terrarium can be very frustrating.

For your tool you can use some type of hooked instrument—tongs, chopsticks, a bent piece of wire.

Start by making a small hole 2 inches deep in the soil where you want your first plant to go. Next, unpot the plant by tapping the edge of the pot against a hard surface, shake all the dirt off the roots and trim them a bit if necessary to fit the opening in the bottle and the hole you have made. Then push the plant down into the bottle as gently as you can. (Don't be afraid if the roots get brushed a bit. Plants are stronger than most of us give them credit for.)

Once you've placed your plant into its selected spot, use your stick to bring the plant to the desired angle, and then tap the dirt down around the plant.

Clean your hands before handling the next plant.

Continue planting until you feel your creation is complete (three or four 2-inch plants are usually plenty for an average 1-gallon terrarium). Remember, they grow rapidly.

Now water the base of each plant with a standard drinking straw half filled. Using the straw will help you to direct the water exactly where you want it to go.

Then add the "frosting"—bits of stone, twigs, anything you feel will help create a complete landscape in a bottle. But whatever you use, be sure to wash it first.

A well-made terrarium can be an artistic delight. Maybe you'll have to experiment with a few, but you'll find that the more you make, the faster you can make them and the more variety you can dream up. Not only is making bottle gardens a great hobby, but they make great gifts!

Just a few more hints on completing your terrarium.

Although many "experts" recommend keeping your terrarium closed or corked, we disagree. Since most bottles have such a small opening and since a cork or a cover allows no air to enter, we've found that corking a bottle increases the humidity too quickly and creates rot where no rot should ever exist. If water builds up on the inside of the glass, it is probably getting too much direct sunlight.

Since it is vital to avoid rot, water your terrarium only about once a month and then, depending on the bottle size, only about a teaspoonful to a gallon bottle.

If a plant should grow out of the top of your terrarium—

and we surely hope it will—gently pull it out, repot it, and put a new, smaller plant in its place.

If, heaven forbid, one of your plants should pass on to that Great Greenhouse in the Sky, remove it from your terrarium with the same type of hooked instrument you used to plant it. Naturally, replace it at once.

Okay. Ready to try one?

Why not call a few friends over and do it together? Terrarium parties are really a lot of fun. But be sure to remind everybody of the first and foremost rule:

Bring Your Own Bottle. ❧

Funky Foliage.

Chapter 9.
Funky Foliage.

Just about everybody has grown a sweet potato at least once in their lives. True?

I mean, don't you remember sticking those little toothpicks in the sides of a yam, setting it in a jar of water, and then marveling at the rapid growth of the long, white roots and the subsequent emergence of tons of lush green foliage?

Well, if you're of such a bent, you can grow many more things in much the same way.

Take an avocado.

First thing you must do—and I admit it's not that easy—is be able to afford one. Presuming that you can, find yourself a good ripe avocado. (The riper the avocado, the sooner the seed will germinate. You might even find, if you come across a real squishy little number, that the seed has already begun to germinate inside the fruit.)

Now halve your avocado, remove the pit, and plant it directly into the soil.

That's it. You read right. Do not pass go, do not collect $200, do not put toothpicks in it and set it in water—just plop the pit right into a pot filled with planting soil, leaving about the top half exposed.

Now then, get yourself a good book and wait.

About two months, usually. So get yourself a good LONG book. (Or better still, a good short book. We recommend *The After-Dinner Gardening Book* by Richard Langer. That book will not only tell you how to grow avocados, but all kinds of funky foliage. You'll learn how to grow the malingering mango. You'll get pointers on pineapples. There's lots of yams, prickly pears, artichokes, pomegranates, papaya plants. You'll even find a chapter called "Yes, We Have No Bananas.")

But back to your avocado. The first sign that your pit has germinated is that it will split.

No, Virginia, I don't mean it'll get up and leave. I mean it'll split in half. And then, my friends, sit back and hold tight.

It's going up, your little avocado tree. It's conceivable—really—that there might come a time when you have to make a critical decision:

Cut a hole in your ceiling or move your tree outside.

When that time comes, you're strictly on your own.

But to me, lazy as I am, it seems a lot easier just to call a

There might come a time when you have to make a critical decision about your avocado tree.

carpenter rather than lug a great big tree outside.

Lots more plants lurk in your grocery bag. Besides sweet potatoes and avocadoes, you can also grow pits of lemons, mangoes, papayas, or litchi nuts.

You can grow a palm tree from the pit of a date.

You can even grow a pineapple from the top of another pineapple.

The full truth of the matter is that you mustn't expect any of your table-scrap plants to bear fruit. It's virtually impossible for them to reproduce, even under optimum indoor conditions.

But they will grow handsomely and sustain for a long, long time as foliage plants if you approximate outside conditions as best you can, give them lots of light, good amounts of water, and feed them a bit more than the usual houseplant. ॐ

Instead of drapes, why not use hanging ferns and Spider plants?

Many people begin their adventure into the world of green simply as a fairly inexpensive method of home decoration. Ever since the first cave-woman put a wrinkled-up leaf in the corner of her cave, people have tried, with varying degrees of success, to bring the outside in.

Naturally there's a limit to the number of plants most people can afford to use in home decoration. That's why it is so important to buy plants that not only look good where you place them, but that will live there as well.

Of course, some people can afford to be extravagant, but alas, most of us have to watch our budgets. We've got to be sure that we decorate with an eye not only to beauty, but with both eyes on survival.

The use of plants in the home need not be limited to a small plant on the coffee table or an African Violet on the windowsill. The concept of decorating with plants can only be limited by your own imagination. If you have a good light situation, look around and start thinking about what you can do with plants.

Instead of filling up your living room with pieces of furniture that do little except collect dust, how about a big Kentia Palm or Dracaena Massangeana for the corner?

Instead of drapes at the window, why not hang some Ferns and Spider Plants? It's really quite simple to put up a "Plant Curtain." Just get some decorative hooks (with toggle bolts if you have a stucco ceiling), several different lengths of chain, a few Ferns and Spider Plants and Grape Ivies and Philodendrons and start hanging.

(I see we've just introduced you to another green friend we haven't talked about yet, the *Spider Plant,* a lovely hanging plant that can be used quite dramatically indoors, especially on a plant stand. Some are variegated and some have dark green leaves. Some varieties send off runners of little plantlets that resemble large spiders. If you like the looks of the runners, make sure the plant you buy has at least one when you get it. Not all have this clever feature. Also called Airplane Plants, they need good indirect light. Water when soil feels dry, mist every day, and feed monthly.)

Instead of a standard room-divider of wood or cement, try Grape Ivy or Kangaroo Vines in macramé holders.

In one of our plant installations we used plants in copper mixing bowls suspended with nylon fishing line (200-pound

Chapter 10.
Decorating
with Plants.

test) at different levels. From a few feet away the plants looked as if they were suspended in mid-air!

When you decorate with plants there are absolutely no rules—other than to have the right plants for the right light. An empty room is an empty canvas. There are no Early American or French Provincial plants. There's no such thing as too many plants in a home—as long as you're willing to make that commitment to take proper care of them.

Planting in decorative containers is another way of using plants for accent. When we first started filling our home with plants, we kept all of them in their regular terra-cotta pots. Then we started gathering up lots of junk—or "junque," at those prices—and began doing lots of transplanting. (We still have our first Mother Earth container in our living room—an old washtub, painted bright yellow, proudly housing a lush, thriving Ficus Benjamina.)

We've put plants in teapots, tin cans, buckets, barrels, even musical instruments. You should see our giant tuba overflowing with Pothos. We call it "Beethoven's First Symphony for Variegated Philodendron." If you're going to plant in a funky old tin or anything metal, we recommend spraying the inside with clear acrylic first. It dries in minutes and prevents rusting.

Putting a drainage hole in the bottom of your container is always a good idea. If the container is metal, that's easy enough to do, and if it's a harder substance, a good drill and masonry bit will usually do the job. Sometimes you won't want to put in a hole, so in that case be sure to put lots of pebbles or cracked-up clay pots in the bottom to absorb water and prevent root rot. In pots without holes, use only plants like Chinese Evergreen, Pothos, Sansevieria, or Nephthytis—hardy plants that need very little water.

Grouping plants is another way to use our little green friends to best advantage. A group can be anywhere from two to two hundred plants in a combination of decorative containers and terra-cotta pots. In groups, plants not only look lusher and fuller, but grow better because increased humidity results from the grouping. (Not only that, they won't be lonely during those hours you're not at home.)

Remember, you are the artist and whatever your eye tells you looks good does look good! And remember this, too—in

We have put plants in teapots, tin cans, buckets, barrels and even musical instruments.

a jungle, there's very little order to the plants; they climb up trees, cover floors and reach for the light—and they'll do the same thing in your home.

Plant stands are good accessories in decorating with plants. Old tables, which can be found at garage sales or swap meets, barrels or small nail kegs turned upside down and painted or varnished, tree trunk stumps, sewer pipes, orange crates, old tripods, even wooden boxes—all make great plant stands. (Right now I'm looking at a Boston Fern named Claudia, sitting fat and sassy on an old Seven-Up crate that we stained.)

White walls are sensational with plants. It's a good thing, too, as most modern apartments have white walls, and landlords are not thrilled when a tenant puts up wallpaper and paints walls. You can find many brackets designed for hanging plants on walls, so give it a try. Some are wrought iron, some wooden. The white of your walls helps reflect the light, a real plus, and you can take your plants with you when you move. Ever try taking wallpaper?

Almost all kitchens have a shelf above the sink or a windowsill where plants can be used. Sometimes special plants like Maidenhair Fern or African Violets can be grown only in the kitchen where there's good light plus additional humidity from either sink or dishwasher. Try shelves across the windows. Fill them with plants and you can have a beautiful indoor garden. And how about a plant on the kitchen table to keep you company over breakfast, maybe planted in an old cup or teapot or coffee can?

You can "furnish" virtually every room in your house with plants, including your bathroom. Actually, since bathrooms—especially those with showers—are the most humid rooms in your home, they are probably the most ideal rooms for plants. Such beauties as Baby's Tears, Irish Lace Ferns and Artillery Plants are generally considered excellent plants for the bathroom. Irish Lace Fern is probably the most graceful and beautiful of all Ferns. Its color is rich green and the fronds are delicate. The foliage is quite dense and at first glance appears to be one green ball. The Artillery Plant is often mistaken for moss and looks a good deal like parsley.

You can "furnish" virtually every room in your house with plants.

All plants love to be in the bathroom, as long as they have light. We live on a busy street in a big city and when we lie in the bathtub surrounded by beautiful living plants, we can pretend we're on a lush tropical isle, and that's fun!

Decorating with plants is creative and interesting and you don't need to be a professional decorator to do a good job. However, a plan is necessary. We recommend drawing your room, marking where your light sources are—and how lucky you are if you happen to have a room with a skylight—then draw in your furniture and finally, sketch in where you want your plants to go. (It's probably going to be necessary to measure some of the areas as it's difficult to look at the width of a tree and guess if it'll fit between the sofa and the chair.) Then, armed with your plan and your budget—off to the plant shop or nursery.

When you get your new plant family home, an old basket, formerly the home of the ironing, can become a home for Grape Ivy. A table formerly covered with long-ago-read magazines and dust becomes its base. The plant is alive and beautiful and will grow happily in a north-window light.

Twin Dracaena Marginatas might make their home in etched terra-cotta pots on the hearth . . . suddenly that end of the room springs to life. The possibilities are becoming realities. On the mantel goes a Philodendron Cordatum. Its former home of green plastic sits in the trash and now the lovely vines are settling into a piece of hand-thrown pottery. A tall, graceful Kentia Palm stands proudly over the chair in the corner, and now the chair doesn't seem to need that upholstering so much. The old nail keg that almost got thrown out a half-dozen times now proudly supports a Neanthe Bella Palm, and with its new, dark-stained paint job looks like it's fresh from a trip to the local decoratory.

Stand back and look around. We were right, weren't we? It's beautiful and you know it. Just remember this one little piece of advice. As beautiful as healthy growing plants are, dead or dying ones are frankly a sore sight for eyes.

Don't leave a sick plant in the room as part of the decor. You can take the little sickies to your plant hospital (Chapter 6) and try to nurse them back, which costs nothing.

Especially if you've got Green Cross insurance. ᕙ

Stand back and look around. It's beautiful and you know it!

Most Bromeliads thrive on partial neglect.

I was first introduced to Bromeliads by a 75-year-old semi-retired mechanic named Rafe "Frenchy" DeLago.

At least I thought I was.

It turns out that I was actually first introduced to Bromeliads by my mother and the Dole Company, but neither my mother nor I knew it at the time. Truth is, my mother still doesn't.

You see, all pineapples are Bromeliads. In fact, all Bromeliads are pineapples!

Confusing? Maybe, but let's simplify it down to this:

Bromeliads are as different from ordinary green plants as cacti or succulents are.

They are a breed apart, and, because he is a certified Bromeliad "freak," so is the aforementioned semi-retired mechanic, Frenchy.

I introduce you to Frenchy here because, my friends, he has written this chapter.

"Frenchy," I asked over the phone, "what are you doing?"

"Making onion soup, dolling," he replied. "Then I'm going to feed my plants."

"With onion soup?"

"Why not?" Frenchy laughed. "If it's good enough for Loretta, it's good enough for them!"

I changed the subject. "Frenchy," I said, "how'd you like to be a writer?"

"A writer? What kind of writer? Maybe some of that pornographic movie stuff?"

"Cut it out," I said, but Frenchy didn't hear me. He was laughing.

"You know there's a place down the street where you can come in and paint a nude girl," he said. "I was thinking about going over to Sears and getting me a roller!" He roared with laughter now.

"Frenchy! I'm serious. I want you to write a chapter on Bromeliads for our book."

Well, Frenchy did—and, God bless him, he's going to come out the star of the show. In his own words, here's your Bromeliad lesson from a genuine expert, Frenchy DeLago of Hollywood, California. . .

Not a bad day . . . 12 gallons of gas . . . two quarts of oil . . . put a patch on a ball for the Johnson kids . . . Mrs.

Robbins from across the street come in with the litter pan of her three cats. Frenchy, it leaks, please fix it, it stinks up the house. Okay. Thank you, Frenchy, have a nice weekend.

Gee, it's 5:30 p.m. I better close up and go home. Furthermore, my old Buick has no lights and I don't want to get a ticket—so let's lock up and scram. . . .

Not so fast . . . in comes Mr. McGonigle . . . Frenchy, wait, my car misses and I want to go Las Vegas. Bring it in Monday, I'm going home for the weekend.

Please, Frenchy, I want to get up there tonight . . .

Sorry, Mr. McGonigle, if you must go to Las Vegas take one of those champagne flights with the Bikini Stewardesses. It's late and I have to go home and feed my Bromeliads.

Bromeliads? What kind of animals are they?

For your information, Mr. McGonigle, Bromeliads are not animals but plants, large ones, small ones. All have beautiful flowers and some you can eat and they come in cans as the pineapple and I've got eight or nine hundred.

Frenchy, I like plants. I have a sweet potato in a mayonnaise jar on the shelf over the kitchen sink and it grows like mad and it's lots of fun, but—

You like plants, Mr. McGonigle? Then come home with me and let me show you my garden. Well, it's not exactly a garden, it's all cement . . . most of the plants are stuck in rocks or nailed or glued on pieces of redwood or logs and the rest are in small 4-inch pots and hang in the air. Come with me, Mr. McGonigle, and I'll give you a couple of Bromeliads to add to your sweet potato patch. . . .

Okay, now let's have a look at what these animals (Bromeliads) are like. Where they come from, where and how they grow, and how they propagate.

The family (Genus of Bromeliaceae) is divided into about twenty different species, such as Ananas, Aechmea, Billbergia, Bromelia, Cryptanthus, Guzmania, Neoregelia, Tillandsia, and Vriesia. (These are the most important for home and apartment gardeners; all of them will grow and bloom for anyone, providing they are in the right location— the plants, that is—have good drainage, lots of air and water and light according to the species.) Bromeliads will grow in pots, in the ground, on trees or rocks or tennis rackets or even on nothing! That's true! You can grow Tillandsia Usnoides (Spanish Moss) right on a coat-hanger!

Got it so far, Mr. McGonigle? Good. Let me show you some of the more common plants and describe their care for you.

Ananas—This is the common pineapple plant. If you buy a pineapple in the store, twist the top off—don't cut it, twist it—then stick it in a can of water for about three weeks and roots will form. Then plant it in good potting mix, keep the soil on the dry side, spray the leaves often, and fertilize every two or three months with Spoonit or any other good all-purpose plant food. In about two or three years you'll have a pineapple—not as large, not as good, and a lot more expensive—but more fun and a conversation piece.

Aechmea—The flower that grows from the center (or cup) of this plant is a beautiful pink and purple, very colorful, and lasts from six to eight months. Needs good light indoors. Keep the cup of your Aechmea filled with water, and the soil barely humid.

Billbergia—One of the most prolific and easiest of all Bromeliads to grow, either in the ground, in pots, in hanging baskets, nailed to boards or logs, on rocks, anything! These are airplants and they get all their nourishment from the air, with the exception of an occasional squirting which you must do. They can take full sun or full shade and they bloom readily all year. The flower only lasts about a week or ten days, but it pups freely and blooms. (Note: A pup is a baby Bromeliad, among other things. All Bromeliads reproduce themselves at the base of the "mother plant," and it is not uncommon to see a small pot filled with as many as eight or nine plants, all in bloom and all around the mother plant.)

Guzmania—This is a medium-sized, very beautiful plant with a colorful large center-flower spike and soft green leaves. It's rather difficult to grow, needs diffused light and high humidity, as in nature it grows in rain forests, so be careful about buying Guzmanias, even though they're sold as houseplants, because they're really greenhouse candidates.

Neoregelia—There are many varieties, large and small, and most can take full sun. All need good light to grow, bloom and take on a good rich color in their centers. Some Neoregelias are nearly black, some are bright red, others red and white striped, some with black or dark red mottlings. A characteristic of the Neoregelias is they all have a small red tip on each leaf and are therefore called the "Red Fingernail

Plants." They are very, very easy to grow inside and they are very prolific. A Neoregelia is really worth having as it is hardy and can stand lots of neglect.

Tillandsia—This Bromeliad is an airplant. It grows in great colonies on trees from Florida to Argentina. Most Tillandsias are quite small, with the exceptions of a few giants such as Tillandsia Prodigosa which grows up to 10 feet tall. Tillandsias flower readily, multiply profusely, and are very hardy. Just spray your Tillandsias twice a week and they'll be very grateful.

Am I boring you, Mr. McGonigle? Okay, let me show you one more.

Vriesia—The subfamily of Vriesias can truly be considered the Royal Family of Bromeliads. The flower spike of Vriesias comes in all colors from a dark red, nearly black, to a pale yellow, and its size is from a few inches up to 8 feet! The "flower" is either single, flat and swordlike, or with multiple flat branches. The leaves may be either unicolored or show fantastic variegations. Vriesias are easy to grow as long as you give them what they need: high humidity, good filtered light, good drainage, and occasional feeding. It's a good idea to move your plants around until you find out where they thrive best.

Generally speaking, Mr. McGonigle, don't overpamper your Bromeliad and absolutely don't overwater. Most Bromeliads thrive on partial neglect—they look dead, but are they? A friend of mine gave me about fifty Aechmeas and Neoregelias he had thrown away as dead; they were all dried out, the centercups had rotted away—they were really a hopeless mess. I cleaned them up and replanted the root stumps, immersed them in a growing solution, repotted them and every one came back in a very short time. Here. Take this Aechmea Fasciata. It's just starting to bloom.

So Mr. McGonigle got started on Bromeliads. He has about 75 beautiful plants now, he's joining the Bromeliad Society, he swaps plants with other Bromeliad freaks, has made lots of new friends—claims that the plants are far better therapy than Las Vegas, but his car still misses.

Bromeliads will reward you with wonderful flowers and colorful leaves. So don't be discouraged if you fail a few times. Remember, the first apple-strudel you baked was lousy but the later ones got better and better. Thank you.🦢

In about two or three years you'll have a pineapple.

Glamour plants are the ones that everybody wants to buy but almost nobody can keep alive. It's really too bad, because such flashy and beautiful plants as the Aphelandra, Gloxinia and Maidenhair Fern would certainly be a wonderful addition to any household planting decor. But like most anything else that's glamorous, they're strictly a luxury. For those of you who want to take a chance and try them, however, here they are, briefly described. Because of the difficulty in keeping them alive, we would suggest you talk their care over with your nurseryman. And take notes. Remember, they are not hassle free.

Anthurium—Native to the tropics, you can see these plants growing wild in Hawaii. However, it's not quite the same in your home. The Anthur has small dark green shiny leaves and is easily recognized when in bloom by a red lily-like flower. The plant needs warmth, humidity and moist soil while in bloom. After the bloom goes away, place it in a cooler spot, let it go on the dry side for about six weeks, then return to warmer spot and begin weekly feeding of fertilizer for blooming plants.

Aphelandra (Zebra Plant)—A beautiful green and white leaf and a brilliant yellow flower spike make this one of the most glamorous plants in the business. Sold all over the country, usually when in bloom, it's really what we call a "florist's shop item," and doesn't do well outside a greenhouse. If you keep your Zebra Plant in a very light place and water it a lot, it might survive for a good long time. But it will always drop its lower leaves after a time and grow into a tall, stalky plant with a small cluster of leaves at the top. Some people like the looks of a freakily-stalked Aphelandra, but most prefer the fresh, full plant. The best advice we can give you is to enjoy it while you can, then cut it back once the flowers die and put it in a nice, bright, warm spot out of sight. Keep it on the dry side and watch. It comes back. Our Aphelandra Billy has done it twice—but no blooms have appeared.

Azalea—My advice on caring for an Azalea is this: Enjoy it while you have it. The blooms can last two to three weeks with constant watering, good humidity and good light. After the blooms disappear, cut the plant back, put outdoors in the shade, give it lots of water and fertilizer, and it just may have a second life as an outdoor plant.

Chapter 12.
Glamour Plants.

Begonias—You will see many different types of Begonias in nurseries and plant boutiques, but with the exception of the new Elatior Begonia, which blooms and lives very nicely indoors for quite a while, most of them really do much better outdoors in a shady area. Paradoxically, the most beautiful Begonias I have ever seen were living happily in a terrarium—a giant brandy snifter with plastic wrap over the top! In other words, in their own greenhouse. You might want to have one in your collection, but just don't expect them to be one of the family for a long while. Keep them in medium filtered light, water just enough to keep the soil slightly moist, make sure the room is cool, and always set them on a dry well. Do not spray Begonias or any other hairy-leafed plants.

Coleus—The Coleus is colorful, grows quickly and is quite beautiful. Always be sure you buy one that has been grown for indoor use; otherwise it will have a very short life span. The hardiest must be grown in a *bright sunny window* and kept moist. Don't mist the Coleus. Misting causes the color to fade from the leaves. For easy propagation, just make a cutting and root in vermiculite or water. Feed regularly, using one-half strength plant food every two weeks. Pinch back new growth to keep plant full and well shaped.

Croton—Brilliantly colored, exciting to look at, many different varieties, thin leaves, fat leaves, small table plants, attractive floor plants 4 feet tall—the Croton! Sounds great—right? But don't be fooled. They are tricky! Crotons must have exceptional light and an environment with good air circulation that's fairly warm. They like to be kept uniformly damp and will droop and lose leaves when they need water. A dry well is necessary. If you have sick Crotons, put them in your brightest window, feed with SUPERthrive, and they can come back. Watch out for red spider. Keep the plant clean by daily misting. Remember, these beauties grow wild outdoors in Hawaii, so try to keep your Croton's home as much like Waikiki as you can!

Cyclamen—Very pretty, but life expectancy same as cut flowers. When in bloom keep cool and humid and soil damp. Fertilize every two weeks. Cut down on water when in bloom.

Gloxinia—What we've said about Cyclamens goes for these Glamour Plants, too.

*If you must have
Crotons, try to keep
your home as much
like Waikiki
as you can.*

Maidenhair Fern (Adiantum)—Oh, I've heard stories about people who raise Maidenhair Fern successfully in their houses, but I've yet to meet one of them. (I'm sure we'll be inundated with mail now from people who have entire mansions filled with nothing but thriving Maidenhair.) But of those of us less fortunate, when we see our beloved Maidenhairs beginning to dry up and turn brown, we must take heart that all is not lost. The fern is not dead! Just wait until it looks like it is, then cut it all the way back down almost to the top of the soil, put it outside in the shade, water it frequently (every day if you want) and sure enough, it'll come back every time! To keep it as long as possible inside, find a nice bright spot, keep it on a saucer filled with pebbles and water, water every day, feed every week and spray with warm water daily.

Well, you're on your own, now.

Good luck with your plants. We know if you honestly get involved as we've suggested, they'll grow for you beyond your wildest dreams.

They certainly did for us.

But please don't expect it all to be a bed of roses. You'll have your losses, just like we've had ours. But they'll be minimal. And something will always pop up during that sad day to brighten things.

Like what happened in our house just this morning:

I was berating our 15-year-old daughter for spending too much money on a dress.

"Money doesn't grow on trees," I barked.

"Yours does," she countered.

Lynn smiled. I shrugged.

And that ended the argument.

Chapter 13.
Now Get
Growing.

"Now get growing."

Index Continued

Index Continued